The Solar Fraud
Why solar energy won't run the world

Howard C. Hayden

VALES LAKE PUBLISHING, LLC

Library of Congress Control Number: 2001097915

Hayden, Howard C.
The Solar Fraud
Howard C. Hayden, Editor
Includes bibliographic references and index
1. Solar Energy
2. Solar Intensity
3. Biomass
4. Wind power
5. Hydropower
6. Solar electricity & Photovoltaics
7. Solar heat

ISBN 0-9714845-0-3

Cover Design by Chris Mascarenas

$21.95 from
Vales Lake Publishing, LLC
P.O. Box 7595
Pueblo West, CO 81007-0595

Dedication

This book is dedicated to those who provide us with energy. A sampling of them includes oil-well drillers, coal miners, nuclear technologists, power-plant operators, energy investors, wind-turbine manufacturers, hydropower workers, lumberjacks, wood splitters, photovoltaic producers, tank-truck drivers, pipeline builders and maintainers, high-voltage power line personnel, utility electricians, Arctic oil-riggers, scientists and engineers who work in the desert to keep solar systems operating, and technicians who work with utility distribution systems to keep the cheapest power on line.

Keep up the good work. Energy is essential for survival, and we appreciate your efforts.

Acknowledgements

Proofreading this book was not an easy task. By good fortune, three family members devoted considerable time to checking numbers, grammar, style, and organization. My wife, Jill Moring, and my two daughters, Alexis A. Hayden and Vanessa C. Hayden, applied their own wizardry from chemistry, astrophysics, psychology, statistics, and technical writing to find the bugs and glitches that crept into the text. I am extremely grateful to have had their help in this project.

Despite their efforts, it is a certainty that errors remain in the text. The blame belongs on the shoulders of the author who put them there in the first place.

Preface

Physics, we are often told, is dry enough to be a fire hazard. Physics underlies absolutely everything about solar energy. One would suppose, therefore, that solar energy must be boring beyond belief.

Nothing could be further from the truth. Solar energy inspires passion, enthusiasm, and devotion.

If sunlight is to supply us with energy we must understand two completely independent subjects: solar energy *per se*, and our consumption of energy. That is, we must understand how we use energy, how much energy we use, and what role the sun might play.

The problem boils down to numbers. Just as it takes a bigger kitchen and more potatoes to feed a regiment than to feed a family, it takes more energy to run a nation than to run a household. Similarly, the solar collectors on a calculator are fine for the purpose, but those little things are obviously inadequate to run a steel mill. One needs to know *how much*.

And what about the title of this book? The word *fraud* implies deceit and/or trickery, and obviously does not apply to solar energy, a topic of science, any more than it would apply to a light switch or a loaf of bread.

Fraud is involved when charlatans sell junk like perpetual-motion machines or little magnets that are advertised to improve fuel efficiency in cars; however that is not the subject of this book.

The solar fraud is the litany of unrealistic, rosy predictions of a solar future. It involves lying with statistics and attempting to manipulate the public through numerous coercive means. It is the sure path to Brownout Nirvana.

Table of Contents

Chapter 1. Introduction

Let's cut to the chase. Energy is the foodstuff of civilization. Energy enables us to modify materials and to move them from place to place. Our Daily Bread would not exist if we did not use energy to till the soil, to grind grain, to move the flour to the bakeries, and to bake the bread.

Energy drives the economy, but it is a serious error to imagine that it's simply a matter of profits. We would need energy to transport goods, even if there were no profit-minded corporations. We would need energy to heat our homes, to pump our clean water, to deliver our milk, and to till the soil, whether or not there was such a thing as money.

We use energy to build not only our homes, but also everything in our homes. In short, energy drives society — every economy everywhere, every *civilization* everywhere. Civilization would simply vanish without energy.

Solar energy plays a role and always has. Modern technology has produced better ways to use solar energy than were available to our ancestors. For just one example, there are western ranches in locations where well water is too deep to be brought to the surface by the old-fashioned multi-bladed wind turbines. They need electric pumps, yet power lines are often very distant; solar power is ideally suited to the task. In this case, the problem can be solved either by solar cells or modern wind turbines equipped with electrical generators.

Scientists and engineers almost universally recognize that solar energy is extremely useful in such niche applications. They also recognize that solar energy — using all conceivable technologies — will not be adequate to run an industrialized world.

Jimmy Carter's Call to Arms

After the OPEC oil embargoes of the 1970s, the price of oil skyrocketed. The price per barrel of oil was about $5.00 in 1971, but rose to $12/bbl by 1974, and further to $25/bbl by 1979. This was a five-fold increase in a mere eight years, equivalent to a compound annual increase of 27%.

On June 29, 1979, President Carter called for a "national commitment to solar energy," with the goal of producing 20% of the nation's energy from various solar sources by the year 2000. The Sunday supplements were filled with stories of solar-heated houses and every conceivable solar energy device. There were learned articles about the economics of solar energy: great demand would result in great competition and lower prices.

Everybody, it seemed, recognized the need for energy supplies to sustain an industrialized giant, and visionaries imagined that vast amounts of energy would come from the sun.

Using Carter's expression, Congress declared the energy crisis to be the moral equivalent of war, while passing the Energy Conservation Act of 1978. Conservation would reduce our energy consumption, especially of oil, and solar energy would step into the breech.

... but Reality Reared Its Ugly Head

The official predictions of a mere two decades ago were wide of the mark. For example, in 1980, California's Public Utility Commission then extrapolated with three curves of 8%, 18%, and 24% annual increases (See Fig. 1 so that by the end of 1988, the price of oil would be $52/bbl, $101/bbl, and $195/bbl respectively.

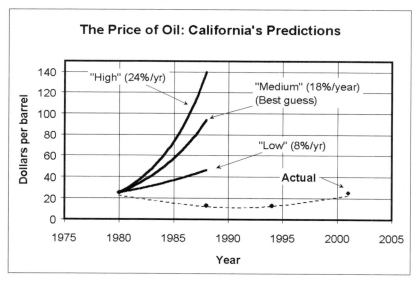

Figure 1: California PUC's 1980 predictions of the future price of oil, compared with actual data.

The facts turned out differently. The laws of supply and demand kicked in; high prices *increased* supply, in turn forcing the price back down. OPEC's lock on the market was soon broken. In 1988, the price of crude was under $13/bbl, and by 1994 was slightly over $13/bbl. California's *lowest* prediction was fully four times as high as the actual price. As of spring 2001, the price has doubled from $13/bbl (1994) to about $25/bbl.

The predictions about solar energy were as bad as the predictions about the price of crude oil. Let us have a look at data from the Energy Information Agency (EIA) of the Department of Energy (DOE), which is the best source of information about energy usage, past and present. EIA's *Annual Energy Outlook 1998* is the source of information for most of the data on energy usage given throughout this book. The US uses about 101 exajoules of energy annually, 36 percent of it to produce electricity (see Fig. 2) at an overall system efficiency of about 35 percent.[1] The electrical energy we use amounts to 12.6 percent of the overall energy budget.

All US Energy

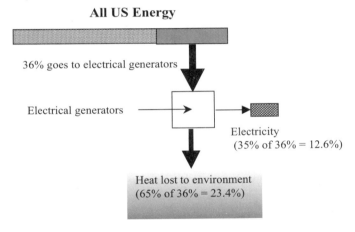

Figure 2: The use of energy to produce electricity in the US. Thirty-five percent of our primary energy is used to produce electricity. The overall system efficiency is 35% . Therefore the electricity we use is 12.6% of the primary energy. Some 23.4% of our primary energy is lost to the environment as heat.

Let us look briefly at the actual solar energy the US produces. The full complement of solar-heat collection equipment built between 1974 and 1997, providing that they all are still in service, for example, produces about 0.02% — one part in 5000 — of our energy consumption. The number of hydroelectric generating stations increased from 3275 in 1980 to 3362 in 1995, but hydro's fraction of our electrical energy *decreased* from 12.1% in

---

[1] See "Efficiency in Heat Engines," p. 29, and "Heat Engines," p. 191.

1980 to 9.1% in 1998. Together, photovoltaics, wind power, wood burning, and waste burning in 1998 resulted in a mere 1.6% of the electricity used in the US. In all uses combined, these four sources amount to 3.36% of the US energy budget; include hydro, and the fraction is 7.3%. This is a far cry from the Carter Administration's call for a 20-percent fraction.

There are two venerable solar sources — biomass (firewood, farm waste, and others) burned for heat, and hydropower. Together they account for almost all of the so-called "renewable" energy in our budget. If we exclude the burning of wood and waste, leaving wind and direct solar energy, and we're down to a contribution of one part out of every 823 — 0.121 percent —— of our electricity. For industries, homes, commercial establishments, and utilities, the total 1998 contribution from wind and direct solar energy (of all kinds) was one part in every 862 —0.116 percent — of the total US energy budget. (For simplicity, we will round both numbers off to one part in 850.)[2]

That is, *technosolar* energy — direct solar heat, photovoltaics, solar-thermal electrical production, and wind — produce about one part out of about every 850 of our total energy and one part out of about every 850 of our electricity.

... and Browner Invokes the Great Hydro Drop-Out

Most people would consider hydropower as renewable energy, because after all, it does rain year after year, putting more water behind the dams. Clinton appointee EPA Head Carol Browner removed it from the list of "renewable" sources. It is renewable, but it isn't "Green," because dams flood the area behind them. Technically, this action reduced our "renewable" electricity by 82%, but only on paper. She didn't succeed in destroying hydropower plants. But the battle isn't over.

In 1979 when Carter made his call to get 20% of our energy from various solar sources, we got 8.47% of our energy from conventional hydropower, geothermal energy, biomass, wind energy and direct solar energy combined. Today, the energy from these politically favored "renewable" sources has *decreased* to 7.3% of our energy.

The Solar Industry

There is a large solar industry in the United States, consisting of manufacturers of wind turbines, photovoltaic arrays, deep-discharge storage

---

[2] Data from *Annual Energy Outlook 1988,* Department of Energy, Energy Information Agency, DOE/EIA-0384(98).

batteries, solar-heat collectors, woodstoves, and many other devices to enable people to collect solar energy in one form or another. Most of their products are not meant for the urban environment where most people live; moreover, the devices are usually cost-effective only where other sources of energy are not available.

Both the manufacturers and the sales personnel recognize the limitations. The owner of a local sales company (near the author's home in Colorado) has prepared a three-page printout telling prospective buyers of the caveats of solar energy. It has a list of *Dos* and *Don'ts*. One item that the blurb does *not* say is to shut off your utility power.

In the words of one wind turbine-factory executive, "I know of no one that has even the slightest thought that renewable will ever replace conventional fuels as the primary energy resource in the United States." The man is speaking for solar-energy experts, not the headline-makers. Ralph Nader, Denis Hayes, The Worldwatch Institute, the Union of Concerned Scientists, or the Environmental Defense Fund, to name a few, who have an entirely different outlook.

Solar Energy in Print

One would hope that every textbook on economics, every treatise on the environment, and every diatribe against population would have entire chapters devoted to energy sources, with the latter two types touting the sun as the ultimate savior for mankind. After all:

> ➢ Energy is necessary for mere survival, and more so for prosperity.
> ➢ Many people have claimed that we are running out of energy.
> ➢ It is inconceivable that man's effect on the environment can be discussed without reference to energy.
> ➢ The official enthusiasm shown for solar energy ought to have evoked countless essays on its merits.
> ➢ And how could anybody even contemplate the world's population without thinking of the energy used?

There are literally hundreds of such books, but most do not even consider the topic of energy important enough to be listed in the index. The occasional reference to energy in economics books is often limited to a few remarks about the *cost* of fuel, admittedly an economic issue, but nothing indicating the central role of energy in civilization. In the environmental and population-control books, energy is regarded as a very limited resource, the use of which is necessarily damaging to the environment. By and large, such books have nothing of value to offer to the discussion of solar energy as a source to support civilization.

Additionally, there are numerous books devoted entirely to some aspect of solar energy, such as home heating. Most of these books offer useful advice, but they do not address solar energy in the broad context of supplying adequate energy for all uses of energy in, say, the US.

Some books, often diatribes against nuclear power, the utilities, or Big Oil, attempt to make a case for solar energy. The present book has quotations from such authors sprinkled throughout the chapters.

The Prize: The Epic Quest for Oil, Money & Power, by Daniel Yergin, is an outstanding exception. Yergin thoroughly understands the role that energy plays in civilization.

The Civilization-Stinks Crowd

Paul and Anne Ehrlich[3] argue strenuously that civilization is harmful to the environment. They gauge the "deleterious impact" of a society on its environment by the impact I, defined as the product of three bad things, Population (P), Affluence (A), and Technology (T). That is, $I = P \times A \times T$.

Formulas require numerical quantities, and only population fits that fundamental algebraic criterion. One thing is clearly intended from their "formula." The good societies are small tribes (population near zero) of impoverished pre-stone-age people, for whom affluence and technology must be somewhere near zero on their undefined scale.

> "The deleterious impact on Earth's life support system of the superrich, inefficient United States is 50 or more times greater than that of an average citizen in a desperately poor nation."
>
> Paul & Anne Ehrlich (1991)

This statement is altogether in contradiction to the statement below, taken from the same source.

> "But in the poorest nations, traditional sources supply 50 to 95 percent of energy use ... Very often fuelwood use is unsustainable, accompanied by deforestation and all the attendant environmental ills. ...
>
> Paul & Anne Ehrlich (1991)

---

[3] *Healing the Planet, Strategies for Resolving the Environmental Crisis* (Center for Conservation Biology, Stanford University, 1991).

In their energy chapter the Ehrlichs switched gears to use per-capita power consumption as a proxy for the impact of a society on the environment. Whoa! In earlier chapters, the impact was *directly* proportional to P. In the energy chapter, it is *inversely* proportional to the P. Oh, well, multiply, divide, who cares?

As I write this paragraph, the news media are reporting on the United Nations AIDS conference, and in particular, about what countries have pledged money to help. Every country that has pledged money is one of the world's energy-intensive civilizations. Not a penny has been committed by any third-world country. Bill Gates alone has donated more money to the cause than all countries in Africa, Asia, and South America combined. The generosity has nothing to do with moral superiority. It is likely that Ghanaians and Cambodians would dearly love to contribute to worthy causes, but they can't.

Only advanced, energy intensive countries can spare the resources, the manpower, and the money to help the poor. When earthquakes, tidal waves, volcanoes, drought, and other natural disasters strike, it is the superrich and allegedly inefficient countries that rush to help.

The "Soft-Energy Path" of Amory Lovins

Energy guru Amory Lovins weighed in with his own predictions in 1977. Figure 3 shows his scenario in which "soft technologies" (a.k.a., solar energy) would amount to about 36% of our energy in the year 2000. By the year 2025, "soft technologies" would give us 100% of our energy. To repeat, the US got 7.3% of its energy from all solar sources combined in 1998.

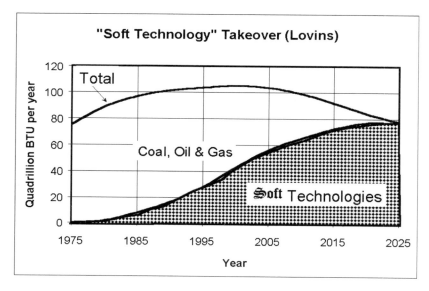

Figure 3: Amory Lovins' "soft-energy path" for the US. By 2000, "soft technologies" (a.k.a. solar energy) would supply about 36% of our energy, and by 2025 they would supply 100%. In 1998, the US got a mere 7.3% of its energy from all solar sources combined.

Post-mortem

Why, one must ask, were the predictions so erroneous? One argument goes this way: the price of petroleum *unfortunately* dropped (read: people *unfortunately* didn't have to pay enough) and made "alternative" energy uneconomical.

> "If oil had remained expensive everything would have fallen into place."
>
> Denis Hayes, 2001

It is odd, isn't it, that anybody should whine that the price of a useful commodity is too *low*.

A related argument asserts that the government had great plans for supporting solar energy, but the Republicans *unfortunately* ended the subsidies and killed solar energy. And why can't a supposedly superior product (solar energy) compete with an inferior product (fossil energy) unless the former is mollycoddled with subsidies? Why, one asks, should a product that ought to generate trillions of dollars in revenue need support from taxpayers' money?

"But perhaps the biggest barrier to deployment of solar technologies is that they are novel."

Paul & Anne Ehrlich (1991)

Americans are known for being averse to novel technology. That explains why ball-point pens, microwave ovens, cellular phones, computers, CD players, television, movies, fax machines, and electric toothbrushes never caught on.

These critics have looked for sociological/political explanations of why sunlight has not become the predominant source of energy. Perhaps they should have looked at sunlight itself.

The Hazards of Following Solar Sirens

Civilization is based on the use of energy, but there is also an element of danger in the story. All energy is dangerous. In the words of Petr Beckmann, to ask for safe energy is to ask for gasoline that doesn't burn. *BUT:*

The most dangerous aspect of energy is not using it.

If the world's citizens were suddenly to stop using coal, oil, natural gas, nuclear power, hydropower, geothermal power, firewood, direct solar heat, photovoltaics, animal labor, and all other sources of energy, there would be no trucks, trains, boats, or animals to deliver food or anything else. There would be no mechanized farming. There would be no clean water delivered to homes. There would be no refrigeration to preserve food, no heat to cook it. There would be neither heat nor refrigeration. Very few people would be around at this time next year to comment about it. People would die by the billions. The massive die-off might warm the hearts of neo-Malthusians, but it would be a human tragedy on an unprecedented scale. Unfortunately, not everybody understands that.

"… we are more endangered by too much energy too soon than by too little too late, for we understand too little the wise use of power;"

Amory Lovins (1977)

Solar energy, as we shall see, is utterly inadequate to meet present needs, let alone to grow as demand increases. To be led down the solar garden-path is thus to be led into dangerous territory. California has taken the first baby steps down that road. Both Sweden and Germany have decided to shut down their nuclear reactors (though not immediately). Sweden's eleven nuclear power plants and Germany's nineteen nuclear power plants produce 46.8 percent and 31.2 percent respectively of those countries' electricity. Even tiny Switzerland has considered shutting down its five reactors that produce 36.0% of its electricity in favor of alternatives.

Some of the leaders of those countries are smart enough to know that solar energy cannot replace the nukes, but the options they offer are in opposition to their own stated positions. Their preferred option is to burn natural gas in combined-cycle power plants. Nuclear power plants do not produce carbon dioxide or any other greenhouse gas; natural gas plants do, regardless of the technology they use. Every one of those countries that have taken the fateful step of deciding on a non-nuclear future has severely criticized the US for not backing the Kyoto Protocol, which demands that the civilized world *reduce* its production of carbon dioxide. (This is double hypocrisy, of course, because they have not ratified the Kyoto Protocol either.)

Solar Sirens beckon us with seductive claims about the abundance of solar energy and politicians allocate tax dollars to expensive solar projects that have no chance of providing sensible amounts of energy. Government agencies coerce utilities to use ratepayers' money to subsidize piddle-power projects, thereby avoiding direct taxation for which they could be justifiably blamed. All-too-comfortable lawyers, politicians, and actors obstruct projects that would provide abundant energy, and coerce the construction of expensive solar toys that can provide precious little energy in its place.

Scientists and engineers worldwide understand full well that solar energy can play only a minor role in the energy picture. The Ehrlichs themselves do a pretty good job of demolishing the solar future. But there is an incessant drumbeat (see Chapter 3) for the world to slouch toward Brownout Nirvana.

Environmentalists

The term *environmentalist* has taken on unfortunate connotations. To some people, the term represents a selfless person who strives to protect the world from sinful, polluting, rapacious corporations. To others, an environmentalist is a pest, an obstructionist, an irresponsible low-life who impedes progress at every turn.

It's unfortunate terminology. In one sense, *everybody* is an environmentalist. *Nobody* likes poisoned water or filthy air.

Still, there is some consistency to the use of the term. In the usual meaning, an environmentalist is a person who rises to object to other people's activities on the grounds of defending the environment. I will use the term in that sense.

The Roadblocks to Solar Energy

There are many critics of solar energy who see the cost as the main roadblock to further development. However, as we shall see, cost is just one of many roadblocks.

One nutty incident came to light in June, 2001. The Stateline Wind Generating Project is under construction along the Washington-Oregon border. Some 450 wind-turbines are supposed to produce 300 megawatts of electricity under full-tilt winds.

Just as a squirrel will often shut down a grid by committing high-voltage suicide on transformer terminals, a ground squirrel threatens to shut down the entire wind-power project. The squirrel is listed as a threatened species — in Oregon. The squirrel is *not* listed as a threatened species in Washington. It seems that a colony of these squirrels has been discovered on the Oregon side of the border, right where they want to put in the wind-turbines. Environmentalists are wringing their hands over a dilemma. Here is one of their favorite children — "Green Energy" — conflicting with another — endangered species.[4]

Some people assert that big-energy politics (see "The Don Quixotes," page 56) is the main roadblock to solar energy development. That is a dubious claim, for many reasons. Big oil companies like Exxon are in the business to make money; if they can make it by selling electricity from big wind farms or big solar installations, they would be perfectly happy to do so.

There are, in my opinion, two main obstacles to solar energy development. They are:

>   The sheer scale of solar projects large enough to accomplish anything useful
>   Environmentalists

In case the reader snoozed, let me emphasize that last point.

---

[4] Jonathon Serrie, "Windmills vs. Squirrels?" *Fox News*, June 28, 2001.

Aside from the sheer scale of solar projects large enough to accomplish anything useful, the main roadblocks will be environmentalists.

"The fiercest battles on the Big Island of Hawaii have been over a controversial proposal to build a 500-megawatt geothermal power plant. ... A pamphlet published by the Pele Defense Fund protested this desecration of Kilauea: 'her person, her body/spirit, her power/mana (spiritual power) are the land of Hawaii."

Berman & O'Connor (1996)

Some of us have wondered why gods like Pele that can make lava flow from mountains are in need of protection from steam pipes and electricity. But Hawaii is not the only place where geothermal power has been opposed on religious grounds. In 1982, under pressure from activists, the New Mexico Public Service Commission killed the (hot-water) geothermal Jemez project in New Mexico on grounds that it would desecrate Indian religious shrines.

But religion is just one reason for opposing renewable energy; the anti-energy crowd has shown remarkable ingenuity in dreaming up reasons to oppose every other kind of energy project. Why not solar as well? Indeed, *most* people, given the scale of solar-energy projects, will object to them on environmental grounds.

Chapter 2. Energy in the US: Brief History

Americans are pictured in the press as energy gluttons in love with fast cars, bright lights, too-comfortable houses, and disposable everything. We have a standard of living that is simultaneously the envy of the world and the target of scorn and derision. Let us take a look at how the US has used energy in the past and is using energy now.

Energy and Power: a Primer

Energy and power are different concepts, just as different as the concepts of where we are and how fast we're going. Unlike location and speed, however, energy and power are very abstract concepts.

There are many kinds of energy, but it is useful to think of it as the heat that you might get by burning a certain amount of fuel. You might express the quantity of energy, for example, as the equivalent of so many gallons of gasoline or so many liters of fuel oil.

Energy can be converted from one form to another. For example, combustion converts the chemical energy in gasoline into heat. However, energy can neither be created nor destroyed. When we speak of "generating energy," we do not mean to imply that energy is created out of nothing. Generating energy is just common parlance for converting energy from one form to a more useful one.

Power is the *rate* at which energy is converted from one form to another. In other words, it tells how fast we're using energy.

For example, a 100-watt bulb converts electrical energy into light and heat, but does so at half the rate as a 200-watt bulb does. A 1200-watt hair dryer converts electrical energy into heat 12 times as fast as a 100-watt bulb. On the other hand, during one day, the 100-watt bulb will use more energy than the hair dryer that is used only briefly, because the light bulb will be left on for hours.

Our monthly bill for electricity charges us for the energy we have used, expressed in *kilowatt-hours*. One kilowatt-hour is 1000 watts multiplied by one hour, or equivalently 100 watts multiplied by 10 hours. It costs about a dime (depending upon where we live) to run a 100-watt bulb for ten hours.[5]

---

[5] Some large industries pay extra if their average power over a 15-minute span is too high, but that is because of the impact on the generating

There is nothing inherently electrical about the watt.   The fuel consumption of human beings, often expressed in Calories[6] per day can also be expressed in watts.   For example, a daily intake of 2000 Calories is equivalent to just under 100 watts.

To distinguish between the two concepts represented by the same unit, it is common to use W_e for electrical watts and W_t for thermal watts.

The Energy Information Agency, the non-political, fact-gathering branch of the Department of Energy, produces tables every year about energy consumption.   Under a heading, "Energy Consumption by Source," they say, for example, that the US used 94.371 quadrillion BTU during 1997, and 94.231 quadrillion BTU during 1998.   The table could equally well have said that the average *power* consumption was that many quadrillion BTU *per year*.   (In Chapter 5, we shall have more to say about these units of measurement.)

The distinction between energy and power (the rate of consumption of energy) is often missed.   For example, Russell Train, former Chair of Carter's White House Council on Environmental Quality, and head (1993) of the World Wildlife Fund compares the rate of consumption (*i.e.*, power) with the amount of resources (*i.e., energy*).

> "With only 5 percent of the world's population, the United States currently consumes nearly 25 percent of the world's energy resources."

> Russell Train (1993)

The writers at *Science News* should know better, but they are in the same boat.   Below, Janet Raloff apparently tries to compare wind *power* with a fixed amount of *energy* (Saudi Arabia's resources.)

> "The U.S. wind power potential, he [Randall Swisher, executive director of the American Wind Energy Association] says, 'is comparable to or larger than Saudi Arabia's energy resources.'"

> Janet Raloff (2001)

Possibly Swisher actually made the error, but Raloff wasn't perceptive enough to catch it.   In all fairness, perhaps Swisher meant something else by "wind power potential."   The amount of energy in the wind during the next

---

equipment.   Otherwise, we simply pay for the electrical energy (in kilowatt-hours), not for the power (in watts).

[6] food Calories are actually *kilo*calories.   1 Cal = 1000 cal.

billion years is certainly greater than the amount of energy in Saudi Arabia's oil.

And finally, some scam artists with a classic pyramid scheme operating on the Internet have a plan to create energy from nothing and to get money from gullible people who can't distinguish between energy and power.

> "We propose to put our generator on your property that will produce 30 kilowatts (kW) *per hour.* The average all-electric home in America uses 2 kW *per hour.*" [emphasis added]
>
> Scam artists on the Internet
> Names withheld to protect the guilty.

Energy History from 1800 to the Present

Let us begin our energy history with an oft-noted but misleading fact: The US uses a lot more energy in a year than we did in the past. Everybody "knows" this, because we have automobiles, trucks, railroads, airplanes, electric lights, household appliances, TV, radio, cellular phones, huge factories, and skyscrapers that our ancestors did not have. Considering all energy sources now and then, our country used about one-hundred-sixty (160) times as much energy in 2000 as our country did in 1800.[7]

But before you run off to destroy your family car, have another look. In 1800, the population was only 5.3 million people, including Thomas Paine, John Adams, and Thomas Jefferson; now it is 275 million, some 52 times as many people. Since these 52 times as many people are annually using 160 times as much energy — and we are including all energy used in homes, factories, transportation, and commerce — then the annual per-capita average energy consumption is (160/52) = 3.1 times as great. In other words you, as a US citizen at the beginning of the 21st Century, use only a little over three times as much energy in a year as did your early 19th-Century ancestors who didn't even have indoor plumbing.

That is, compared to the US of 1800, we have 52 times as many people, each using 3.1 times as much energy. *If* there is a problem with the US using "too much" energy, the great majority of the blame should be

---

[7] Energy data for early years from Chauncey Starr, "Energy and Power," article in *Scientific American* offprint book, *Energy and Power,* W.H. Freeman (1971). Later data from EIA tables. Per-capita energy consumption was about 90 kWh/day in 1800, equivalent to 3.7 kW.

attributed to the population increase, *not* to anybody's profligate use of energy to drive his SUV.

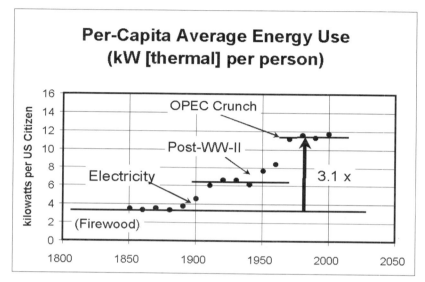

Figure 4: The average energy consumption rate for the US, on a per-capita basis. The average consumption was about 3.7 kW_t per person until the 1890s when electricity and widespread rail traffic increased consumption to around 6.4 kW_t/person. Since the first OPEC embargo of 1973, per-capita consumption has been nearly constant at about 11.5 kW_t/person.

Figure 4 shows the per-capita rate of energy consumption in the US since 1850, expressed in kilowatts [thermal] per person.[8]

The pre-1850 era was characterized by relatively constant consumption from the predominant energy source in use, firewood. Coal, used primarily for smelting, took on wider use as fuel for railroads, and took on other roles as well. After 1850, the use of coal and wood both increased, but the use of wood declined after about 1880 (See also Fig. 6, page 19). Coal became the dominant energy source after about 1880, and remained so until the

---

[8] Example: in 1950, the US population was 151 million, and the energy consumption during that year was 34.64 quadrillion BTU. This is equivalent to 7666 watts [thermal] per person. (The conversions will be discussed in Chapter 5 and in Appendix A.

petroleum era. From the pre-depression era until the post-WW-II era, *per-capita* consumption was relatively constant at about 6.4 kW$_t$ per person. The post-war industrial boom saw an increase to about 11.5 kW$_t$ per person, a value that has changed little in the last three decades.

To gain a little perspective on these numbers, the pre-1850 value corresponds to thirty-five 100-watt bulbs running continuously for every man, woman, and child in the US. The present value corresponds to one-hundred-fifteen 100-watt bulbs continuously burning for everybody in the US.

Human food consumption is equivalent to about 100 watts [thermal] per person. This figure is routinely used in the heating & air conditioning trade to account for the heating of the room by its occupants (in an auditorium, for example). An athlete in excellent shape, but given a more substantial diet, can actually perform work at the rate of about 100 watts. If he works for ten hours, he has done one kilowatt-hour of work. If he is paid $6.00 per hour, the kilowatt-hour of work comes at a cost of $60. The utilities that everybody likes to complain about charge about a dime for the same amount of energy. Scoundrels! They're taking away jobs!

The average per-capita US energy consumption rate of 11.5 kW$_t$ corresponds to the work of 115 athletic servants working for each of us around the clock. Readily available energy puts slave-owners out of business.

Energy in Colonial America

The main source of energy in Colonial America was firewood, obtained from abundant forests. Figure 4 shows only the contributions from firewood, from which our ancestors obtained far more energy than from either their water wheels or windmills.

It is very difficult to assess the energy contribution from animals, partly because we may not agree on the definition. We could regard the energy contribution as being equal to the work done by a single farm horse in a year. On the other hand, we have to feed the horse every day, regardless of whether it is working. Perhaps, in figuring the energy consumption of our ancestors, we should regard the fuel-equivalent of the horse feed. Regardless, when investigators have estimated energy usage by our ancestors, they have not taken animal power into consideration in the first place, let alone the energy in the feed that kept the animals alive all year. That is, the estimates of our ancestors' energy usage are *under*estimates, and therefore the 3.1-times increase in *per-capita* consumption is probably an *over*estimate.

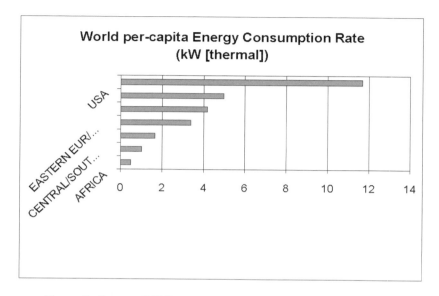

Figure 5: Present (1998) per-capita energy consumption rate. The Middle East presently has the per-capita consumption that Colonial America did. Central and South America, the Far East (including Oceania) and Africa are still below that rate, probably because their warm climates do not require domestic heating.

Even by today's world standards, Colonial America was a relatively advanced culture. Our ancestors used energy to smelt iron, to forge tools, to make copper, brass, and bronze, and to make glass. And, of course, the harsh winters in New England, upstate New York, and much of Pennsylvania required the use of firewood for home heating. Colonial Americans used far more energy per capita than people in Central and South America, the Far East, and Africa use today (Fig. 5).

Firewood in US History

Figure 6 compares the annual energy consumption of the US from 1850 until 1985 with the energy from firewood alone. The unit used there is the *exajoule* (abbr. EJ), which is a billion-billion joules (billion-billion watt-seconds). That is, 1 EJ = 10^{18} (1 followed by 18 zeroes) joules. One EJ per year is about 32 billion watts (either electrical or thermal).

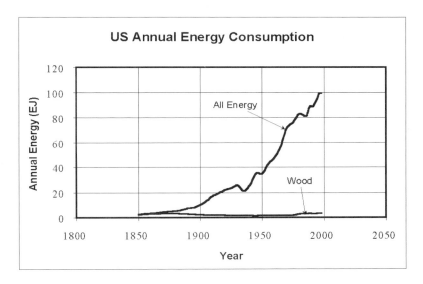

Figure 6: Historical use of energy use in the US. The upper curve shows the total annual energy consumption; the lower curve represents the energy obtained from firewood. The total energy consumption in 1850 was 2.54 EJ and that in 1985 was 76 EJ. [See note 7 for reference.]

Firewood was the major contributor to our energy from the earliest days of the Republic, with coal running a distant second. The wood was used not only for heating and cooking, but also for various small industries, such as glass making. Some coal and coke (made from coal) were used for metallurgy. In the mid-1800s, coal started to be used for heating and for railroad and river transportation. The all-time high consumption of wood occurred in about 1880, when the US used about 3 EJ per year, and the nation's forests were being depleted.

New England today is a veritable jungle, and its few clearings are prized pieces of land. It is hard for visitors to New England today to visualize, but in the mid-1800s, the region had very few trees. The original trees were cut down to provide firewood, to provide lumber, and to clear the land for farming. The land could never have reforested itself if wood had remained the primary energy source.

The current US consumption of all biomass, firewood, farm waste, lumber waste, methane from landfills, cow manure, and all others put together is 3.2 EJ/year, not appreciably different from the value of 1880.

Present Sources of US Energy

The US uses about 101 exajoules (EJ) per year, amounting to 3.2 trillion watts [thermal], or 3.2 terawatts (TW_t). Figure 7 shows the sources of energy in the US (1998). Fossil fuels — coal, oil, and natural gas — account for 84.9% of our energy. Nuclear fission accounts for 7.6% of our energy, and hydro for 3.8%. Biomass accounts for 3.2%.

Hydropower and biomass are by far the greatest *solar* contributors to the overall energy picture, contributing 7% of our energy. The high-technology solar contribution — photovoltaics, direct solar heat, solar-thermal-electric, and wind together account for 0.117% of our energy. That is, the exotic solar stuff —*technosolar* — that people proffer as the up-and-coming energy sources accounts for only one unit out of every 850 units of energy.

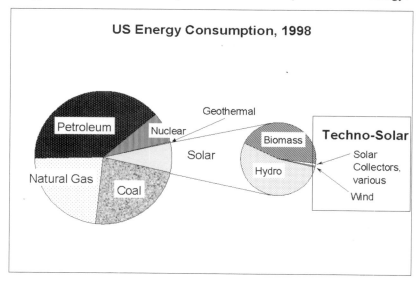

Figure 7: Sources of energy currently used in the US. Petroleum 38.8%; Natural gas 23.2%; Coal 22.9%; Nuclear 7.6%; Hydro 3.8%; Biomass (firewood, waste, …) 3.2%; Geothermal 0.33%; Direct solar 0.078%; Wind 0.038%. The total of all US energy consumption in 1999 was 101.9 EJ, about 27% of worldwide energy use. [Data from EIA tables.]

Referring to 1988, Berman and O'Connor confirm a minuscule 0.07 quads (0.09 percent) of US energy (1988) derived from high-tech solar sources.

"Of a total of 82 quads (quadrillion Btus) of energy consumed in the United States in 1988, only 0.05 quads were derived from solar-thermal plants and 0.02 quads from wind generators."

Berman & O'Connor (1996)

Sources of Energy for US Electricity

About 36% of our energy (36.6 EJ/year, 1.16 TW_t) goes toward the production of electricity. The sources of electrical energy are shown in Fig. 8. Stationary machines (such as power plants) use many sources of energy, leaving petroleum for transportation applications. (Nobody has yet developed a hydro-powered airplane.)

Utilities generate most of the electricity used in the US, some 88.75%. Non-utilities, primarily factories that need huge amounts of electricity, generate the remaining 11.25% of the electricity. The figures shown in Fig. 8 are percentages of all electricity combined. For example, nuclear power generates 21% of electricity produced by utilities, but 18.6% of all electricity combined.

Of interest is the electricity from renewable sources. Hydropower and biomass account for 8.4% of our electricity (*versus* 18.6% for nuclear). All of the high-tech sources — wind turbines, solar-thermal-electric and photovoltaics — account for one part out of every 850 — 0.12 percent of our electricity.

Overall, in 1998, wind turbines in the United States produced 3.5 billion kWh, equivalent to around-the-clock average power of 399 MWe. The wind turbine capacity was 1700 MWe. The average *capacity factor* was therefore 399/1700, or 23.5%.

Electrical Energy Generation

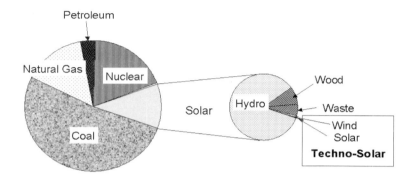

Figure 8: Electricity production from various sources in the US. Coal 51.9%; Natural gas 14.7%; Nuclear 18.7%; Hydro 9.1%; Petroleum 3.6%; Wood 0.93%; Waste 0.58%; Geothermal 0.39%; Wind 0.10%; Photovoltaics and solar-thermal-electric 0.02%. The net electricity production of utilities in 1999 was 3183 billion kWh, equal to 11.5 EJ. Non-utility producers generated another 140 billion kWh, equal to 0.5 EJ in 1999.

Heat from the Earth

Geothermal energy generates 0.38%— one part out of every 260 — of our electricity. It is renewable in some sense, and may eventually become a big player in the energy field. When heat is extracted from a geothermal site to produce steam for a turbine, the site cools down in response. Eventually, geothermal sites become too cool to be of use. That is, any given geothermal site can be considered as a *non*-renewable source of energy. On the other hand, the supply of energy deep in the earth is so enormous that it could last mankind forever — *if* we can develop the technology, and *if* society will tolerate frequent ventures deep into the earth.

Most of the geothermal electricity is produced in California. Those sites are not producing as well now as they did a few years ago.[9] For example, the Geysers Geothermal Power Plant in California was designed to produce 1984 MWe, but never quite achieved the goal. According to Pacific Gas & Electric, "the geothermal fields have been in gradual decline for several years." PG&E expects the capacity to drop to 700 MWe, and is no longer able to use it to supply "baseload," constant, around-the-clock electrical power.

> "In this case, mismanagement can lead to depletion of the local [geothermal] heat source."
>
> Brown, Flavin & Postel (1991)

Mismanagement can also cause water to leave a glass while you're drinking from it.

How We Use Energy

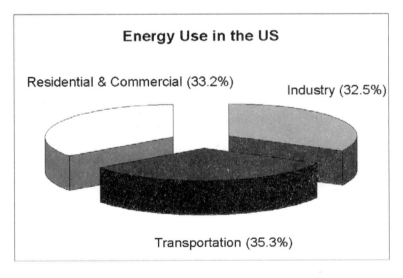

Energy Use in the US

Residential & Commercial (33.2%)

Industry (32.5%)

Transportation (35.3%)

Figure 9: Energy use in the US. [Data from EIA. See note 2.]

---

[9] www.pgedivest.com/eir/chapters/04-08egy.htm

There are about 100 million housing units in the US. Some 50 million homes are now heated by natural gas and 30 million are heated by electricity, as shown in Fig 10. Fuel oil heats another 10 million.

The transportation, industrial, and residential/commercial sectors consume approximately equal amounts of energy, as shown in Fig. 9. Of course, transportation is powered almost exclusively by petroleum. Not shown in Fig. 9 is the electrical industry, because it is a sort of pass-through technology, as we saw in Fig. 2 in Chapter 1. The electrical industry uses 36% of the primary energy to produce the electricity, which is then sold to residential, commercial, and industrial customers. Despite the existence of subways and city commuter railroads, extremely little electricity goes into transportation.

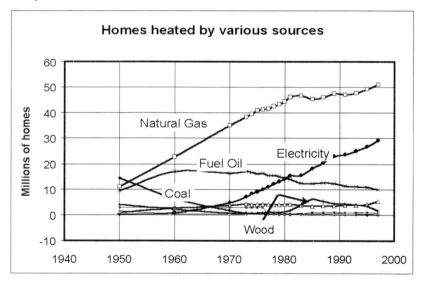

Figure 10: The number of homes heated by various methods, 1950–1997. During the half-century shown, the number of homes heated by natural gas has steadily increased from 10 million to 50 million, while those heated by coal has decreased from 14.5 million to 180,000. Nearly 30 million homes are now heated by electricity. The renewable contributions are shown in Fig. 11 on a much-expanded scale.

Some homes are also heated by alternative sources, as shown by the lower curves in Fig. 10 and the four curves in Fig. 11. Following the oil embargoes of the 1970s, there was a big push to heat homes with firewood.

Of course, that requires either a woodstove or a fireplace, neither of which was particularly common in the 1970s. As factories started churning out woodstoves, firewood heated more homes, reaching a peak of 6.3 million homes in the mid-80s. The subsequent decrease to fewer than 2 million homes heated by firewood was due to both the low cost of fossil fuels and the hard work associated with feeding a woodstove. Notice in Fig. 11 that the number of homes with no heat source at all exceeds the number of solar-heated homes.

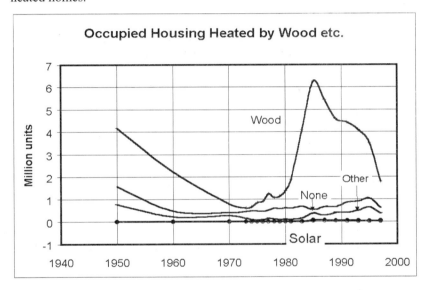

Figure 11: Data from Fig. 10, showing only homes heated by wood, nothing at all, other, and solar. The large increase in wood-heated homes just after 1980 undoubtedly came as a result of the oil embargoes of the 1970s; the *de*crease after 1985 was undoubtedly caused by the realization that feeding a woodstove involves a lot of hard work. Despite the articles in Sunday supplements, solar homes are few and far between. A mere 0.03% of all US homes — one home out of every 3330 — is heated by solar collectors.

Efficiency

Efficiency in Households

The energy used by a typical US household currently amounts to about 3500 joules for every second of the year (See Fig. 12). In other words, the average power consumption in US households is about 3500 watts [thermal].

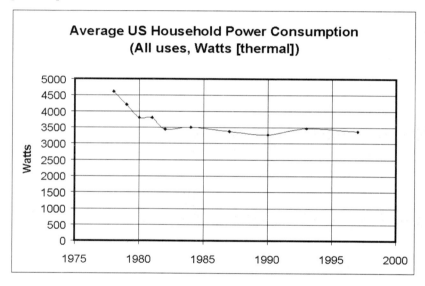

Figure 12: The year-round average power consumption of American households, 1975–1997. (Includes electrical power, heating, and all other uses.)

By comparison, homes in the early 1800s used much more power, mostly for heating. After all, their per-capita consumption was 3500 (thermal) watts (Fig. 4), and families were large. We can make a crude estimate. Assuming that half the energy of the society was used in homes and that the average family had six persons, the family home would have consumed about 10,000 watts,[10] mostly as heat from firewood. The houses

---

[10] Again, this is not electrical watts. It is the annual energy consumption in joules (almost entirely for heating) divided by the number of seconds in a year.

of the era were large and poorly insulated. Moreover, they mostly used fireplaces, which are not nearly as efficient as woodstoves.

Figure 12 shows the average US household power consumption since 1975, showing a gradual decline. The decline has, in fact, been continuous since colonial times.

Efficiency in Automobiles

The pre-war standard for US cars was 20 miles per gallon (8.4 km/liter). The post-war automobile era saw a dramatic decrease in fuel efficiency as designers added a host of energy-consuming features to automobiles. Mostly, however, the cars steadily grew in weight. Many were about as boxy as 30s-era cars, not having been designed for aerodynamic properties. However, the post-war cars were driven a lot faster. Wind resistance causes cars to gobble up fuel.

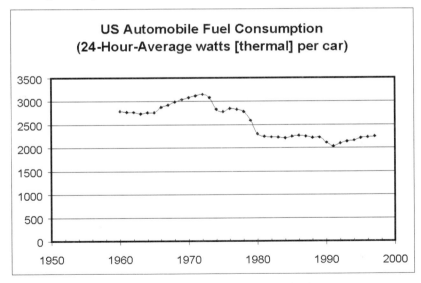

Figure 13: The year-round average power consumption of American cars in watts [thermal] per car.

Following the Arab oil embargoes, cars actually began getting *poorer* mileage for a while, and the cause was pollution-control measures that were being required at the same time as the energy crisis was in progress. For a quick fix, engineers threw energy at the problem, so to speak, in order to clean the exhaust.

Hope was on the horizon. Japanese cars already met some important requirements for the American market. They were smaller and more aerodynamic, and they also burned fuel more efficiently.

The computer has greatly improved matters. In the first place, engineers use computers to design cars and test their performance without having to build them. Secondly, on-board computers also improve the engine performance. They control the fuel input, second by second, looking at the exhaust to assure that the fuel/air mixture is right. They also control spark timing. Moreover, electronic ignition systems have vastly superior performance and durability than the old Kettering ignition system that cars used for the better part of a century.

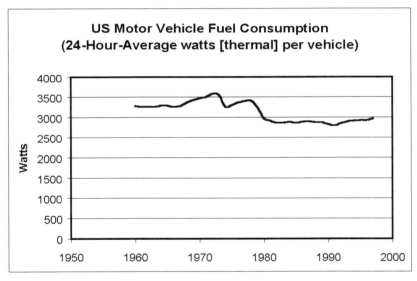

Figure 14: The average power consumption of vehicles on American roads, including motorcycles, buses, other 2-axle, 4-tire vehicles (including vans, minivans, pickup trucks, and sport-utility vehicles), single-unit trucks with six or more tires, and combination trucks.

The [thermal] power consumption of a car in motion is huge, typically in the range of 50,000 W_t (tiny cars) to 200,000 W_t (gas guzzlers) for cars going at highway speeds. On an around-the-clock basis — total annual energy input divided by the number of seconds in a year — things are different. Figure 13 shows that the yearly average power consumption for US cars from 1960 to the present dropped from 2,700 W_t to 2,200 W_t, an 18 percent improvement. Again, for comparison, the 2200 W_t is equivalent to burning twenty-two 100-watt bulbs continuously, every second of every day. The drop in average fuel consumption, however, is completely overshadowed by the huge increase in the number of cars on the road since the 60s.

The fuel consumption data in Fig. 13 include passenger cars and motorcycles. It might be argued that Fig. 13 is misleading, since sport-utility vehicles and minivans are not included. Therefore, we need to look at the average fuel consumption of *all* vehicles considered together.

Figure 14 shows that the average power consumption for all vehicles in the US is about 30 percent higher than that for automobiles alone, but the graph includes everything from econo-boxes to tandem tractor-trailers. There is a slow downward trend in power consumption, as vehicles have become more fuel-efficient across the board.

Engine design is not the only consideration. Jack Stevens, Emeritus Professor of Civil Engineering at the University of Connecticut[11] was involved with a project a half-century ago to change the design of long-haul trucks to make them more aerodynamic. It was clear from the outset that better fuel economy could be achieved by merely changing the shape of the truck. The resistance to change came from truck drivers who thought there was something "sissy" about the new shape. After the price increases incurred as a result of the OPEC oil embargoes, they began to change their minds. The newer trucks are much sleeker and more fuel efficient.

Efficiency in Heat Engines

There is an asymmetry in the law of conservation of energy. Every bit of electrical energy that goes into an oven gets converted into heat, but you cannot take that heat and convert it all back into electrical energy. More formally, you can convert work to heat with 100 percent efficiency. You can convert heat to work, but the efficiency — the ratio of work accomplished to heat expended — will always be less than 100 percent. It is because of this asymmetry that it is necessary to distinguish between watts [thermal] W_t and watts [electrical] (W_e).

A heat engine is a device that converts heat to work. Heat engines go back in history as far as Hero of Alexandria, who made a toy that spun when powered by steam. The first useful engine was made by Newcomen in the mid-1700s.

The Ford Museum in Dearborn, Michigan has the world's oldest Newcomen steam engine on display. The engine was used for pumping

---

[11] Private communication.

water from a mine in Norbury, England. It was removed from service in 1764 to make way for a more powerful engine, but moved to Fairbottom Valley in Lancashire, where it was used until 1827. (Data from Ford Museum displays and from *Henry's Attic.*[12]) The cylinder of the Newcomen engine was 0.71 m in diameter, and the length of the cylinder was 2.4 m. Overall (except for the steam boiler) the engine was 8.5 m long, 5.5 m high, and about 3m wide.[13] I can only hazard a guess as to the weight, but it was probably ten to twenty tonnes.

The power output of the Newcomen engine is estimated to have been about 9200 W_t, about the same as the output power of the Volkswagen engines in the 50s. The efficiency of the Newcomen engine was a mere 0.05%.

James Watt patented a device for removing spent steam from the engine, leading to a big improvement in efficiency. Watt's demands for money for use of his patent kept many potential users from using more efficient engines. (Coal was cheap.) After Watt's patents ran out, engines began to improve. The Ford Museum has a "walking-beam" engine (also huge) from the early 1800s that produced 9000 W_t (12 horsepower), but at 4% efficiency, an eighty-fold improvement over the Newcomen engine.

By 1848, George H. Corliss had produced a 224,000-W_t (300-HP) steam engine of approximately 20% efficiency. In the 1890s, electrical generating plants were using steam for motive power to turn the shafts of generators. The steam engines were still running at about 20% efficiency, but the electrical generators converted 95% of the drive-shaft power into electrical power. The *thermodynamic efficiency* is, be definition, the efficiency of converting heat to work.

The conversion of *heat* to work is notoriously inefficient. The chair you are sitting in has lots of thermal energy owing to molecules vibrating back and forth. However, the only way to make use of that energy is to let the heat flow toward something that is colder. Heat engines always receive heat from something that is hot, but they must reject heat into something that is colder.

In a gasoline engine, for example, most of the heat rejection takes place while the piston shoves hot exhaust out the exhaust valve. Of course, the exhaust must be hotter than the surroundings. The colder the surroundings (all other things being equal), the more efficient our engine could be. If it were possible for our surroundings to be at absolute zero (–273 °C, –

---

[12] *Henry's Attic,* Ford R. Bryan (Ford Books, Dearborn, MI, 1995)
[13] 28-inch diameter, 8-foot length, 28-foot overall length, 18-foot height, 10-foot width, 10-20-ton weight, 11-hoursepower output.

459 °F), it would be possible to convert *all* of the heat to work. But our surroundings are not that cold. Therefore, the efficiency of heat engines is limited. [Another problem is that combustion is poor when the fuel/air mixture is cold; otherwise the fuel economy would be much greater in Alaska than in Arizona.] On the upper end, the higher the temperature of the source, the more efficient our heat engine can be. The limitation here is one of materials. Hot steam is extremely corrosive. As metallurgists improve their craft, the result is improvements in engine efficiency. The very best efficiency we have achieved so far is 60% for a combined-cycle, natural-gas powered turbine whose blades are made of single-crystal titanium. Automobile engines typically have about 25% efficiency.

The historical information about Newcomen engines in the previous paragraphs comes from *Henry's Attic* though it is not clear whether the audio/visual experts at The Ford Museum ever read any of it. Near their beautifully restored historic engines there is a short videotape asserting that we need to improve efficiency. It also complains that the changes (the ones that raised humans from depressing life-long drudgery to present conditions) "did not reflect environmental costs," and wonders hopefully, "can the government promote conservation?" Somehow, they failed to notice the 400-fold increase in efficiency in heat engines in their own display of engines dating from roughly 1750 to 1900, or that drive-shaft efficiency of electrical generators was already 95% nearly a century ago.

To get some idea of the improvement of engines that has occurred, imagine an engine of 0.05% efficiency — like Newcomen's first engine — in one of today's cars. Rather than burning one gallon of gasoline to go 30 miles, it would use about 17 gallons of gasoline to go one mile.

Every machine that converts heat to work is subject to the inefficiency dictated by the laws of physics. (We discuss this matter in more detail in Appendix B.) But if we have a source of heat and we want to do work, we have to use heat engines.

> "For [Amory] Lovins, splitting the atom at a temperature of thousands of degrees to generate electricity (much of which will be lost in transmission) in order to raise the temperature of a house by 30 degrees in midwinter made about as much sense as cutting butter with a chainsaw."
>
> Berman & O'Connor (1996)

Amory Lovins is obviously miffed about the low efficiency of converting heat from nuclear fission to electricity. If either Lovins or Berman and O'Connor can devise a way to use the energy of splitting atoms *without* converting heat to work, scientists and engineers around the world would nominate them for the coveted Nobel Prize for inventing the method.

Efficiency in Generators & Transformers

The conversion of mechanical energy to electricity is relatively straightforward. An efficiency of 95 percent was already common early in the 20th Century. The first large-scale hydroelectricity came from Niagara Falls in the 1890s, already at 85% overall efficiency (water turbines included). Only football coaches and deodorant manufacturers are opting for 110% efficiency.

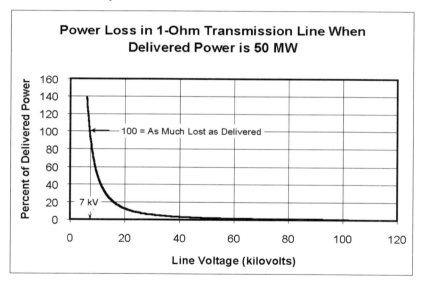

Figure 15: The power loss in a hypothetical power line connecting a generator to a town that uses 50 MWe of electrical power. As the line voltage increases, the power loss decreases dramatically. At 7 kV line voltage as much power is lost as is delivered to the town. If the line voltage is 100 kV, the power loss in the line is only 0.5 percent of the 50 MWe delivered.

Power losses occur in transmission lines between the power plant and consumers. The method for transmitting electrical power over long distances with minimal power loss was due to the eccentric Serbian genius,

Nikola Tesla. A numerical example should suffice to show how it works. Suppose a small town uses 50 MWe of electricity. Suppose further that the power comes from a distant station over power lines that have one-ohm resistance. The power loss depends upon the line voltage, as shown in Fig. 15. For this hypothetical case, if the power were delivered at 120 volts, the power loss in the lines would be 173 billion watts, about 3500 times as much power as would be delivered.[14] If, instead, the power is delivered at 120,000 volts, the power loss in the lines is 173 thousand watts, a million times smaller, and about 0.35% of the delivered power.

The power loss is greatly reduced by transmitting the power at very high voltage; however high voltage is dangerous to the consumer. Tesla's scheme was to generate AC electricity, instead of the DC electricity favored by Edison. The voltage of AC electricity can be raised or lowered at will through the use of transformers, which are nothing more than coils of wire wrapped around an iron core.

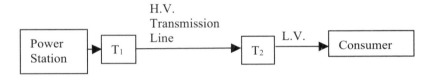

Figure 16: The use of a step-up transformer (T_1) to raise the voltage from the power station for long-distance transmission. Transformer T_2 lowers the voltage for use by the consumer.

Utilities therefore use step-up transformers at the power station to send high-voltage AC over the transmission lines. Step-down transformers (in stages) gradually reduce the voltage until near the customer the last transformer provides the required 120 volts. The transformers themselves are often 99% efficient.

"…the superrich, inefficient United States…"

Paul & Anne Ehrlich (1991)

As one example, LUZ International, builder of the Solar Energy Generating System (SEGS) in the Mojave Desert, built a 220,000-volt

---

[14] The current in the lines would be 50 million watts divided by 120 volts, or 417,000 amperes. The power loss in the lines, I^2R, would be 173 billion watts.

power line to carry electricity from one site to another near the utility grid, when the distance was only 19.3 km, a mere 12 miles. The purpose of the costly project was obviously to achieve high efficiency.

Efficiency in Wind Turbines

Technically, wind*mills* are wind-driven mills for grinding grain. The "windmills" seen by the thousands on American farms were used for pumping water out of shallow wells. Wind *turbines* are generic wind-driven rotary machines. (Never use two syllables where three will do.) Most modern wind turbines are used for producing electricity.

People have used windmills for millennia, but high efficiency has been achieved only since the aeronautical age, especially since propeller research was done at NACA (the precursor to NASA) in the years prior to World War II. Having a wind turbine drive an electrical generator is an obvious application. The technology of hooking a wind turbine safely and efficiently to a power line is relatively recent. Still, everything has been in place for decades, yet we get very little power from wind turbines.

There is a theoretical upper limit of 59% to the efficiency of wind turbines. That is, no wind turbine can be built that can convert more than 59% of the wind's energy into mechanical motion (or to electricity). The millions of windmills that were used on American farms to pump water out of the ground were about 25-30 percent efficient. Today's large industrial-strength wind turbines can achieve almost 50% efficiency under the right conditions. These machines are a far cry from the windmills that became a national symbol for the Dutch. The topic will be addressed more fully in Chapter 9.

Reliability

Reliability has always been important to the telephone companies, who regularly use batteries and backup power plants to provide power when the power line goes dead. It has become a huge issue with everything from hospitals to Internet hotels that host thousands of web sites. As of the present date, the amount of stand-by capacity is around 70,000 MWe (See footnote 15), some 9% as great as the electrical capacity of the US.

The requirements for power are becoming more and more stringent with the digital economy. A brief failure in power at an "Internet Hotel," a building that houses thousands of computers used as web hosts can cost millions of dollars in lost revenue.

The quality of the power is just as important. As Peter Huber and Mark Mills say[15], "Some years ago, a Stanford computer center found its power fatally polluted by an arc furnace over a hundred miles away." The "pollution" was what engineers lovingly call noise on the power line — voltage fluctuations of all kinds. But noise can come from poor sources just as easily as it can from heavy users like arc furnaces.

Solar power, to be of use, must therefore prove itself both reliable and devoid of electrical noise.

The Role of Solar Energy Internationally

The EIA's *Annual Energy Outlook 1998*, Table 11-2 shows the World Primary Energy Production from various sources. Biomass does not have a separate listing because there is simply no way to know how much firewood and buffalo chips are being burned in many energy-starved countries throughout the world. On the other hand, EIA's *International Energy Outlook 2001* says that the worldwide energy fraction due to solar sources is presently 9%, not greatly different from the 7% fraction in the US.

> "Wind and photovoltaics are the fastest growing sources worldwide, and *wind alone*, now the cheapest new generating option in the United States, *is adding more global capacity each year than nuclear power did through the 1990s*."
> [emphasis added]
>
> Amory Lovins (2001)

What's wrong with this quote? In the EIA's tables for international energy, solar and wind are not large enough to be considered separately, so their contributions are listed in their "Geothermal and Other" column. From 1900 to 1997, the total production of energy in that column increased by 0.41 EJ/year. In that same time interval, nuclear energy generation increased by 3.79 EJ/year. Granted that production is a different thing from capacity, it is clear that annual nuclear energy increased over nine times as much as the *combined* increases of geothermal, photovoltaics, wind, and biomass, including chicken manure. In other words, wind is *not* adding more capacity each year than did nuclear power throughout the 1990s.

Denmark is often cited as a country where wind power is making great inroads. The country makes a lot of money *selling* wind turbines; in fact, it

---

[15] Peter Huber and Mark Mills, *Digital Power: Processors to Power Plants* (Gilder Publishing Company, Great Barrington, MA, 2001)

is one-third of their export market. They have plans to get 20% of their electricity from wind turbines within the next decade or so.

The Danish experience, however, is not one of free-market economics.

> "...because wind turbine income is exempt from income taxes, a particularly important consideration in Denmark, where high income tax rates make 1.00 krone of wind-turbine income worth 1.46 krones of taxable income."
>
> Berman & O'Connor (1996)

The news of July, 2001, is that the entire economy of Denmark is smaller than that of Atlanta, Georgia. What works somewhat well in a tiny economy on a peninsula with good onshore winds is not necessarily the correct prescription for an industrialized giant.

Worldwide, the major contributors to solar energy are biomass and hydropower, just as they are in the US. Worldwide, hydropower contributes 6.6% of the worldwide total of 402 EJ/year of primary energy used annually.

Chapter 3. The Solar Drumbeat

Back when people burned firewood for heat, whale oil for lighting, and incense to cover the odor of horse manure in the streets, there were no newspaper articles screaming for solar energy. They were *using* solar energy, one-hundred percent. Now that all solar sources combined provide about 7 percent of our energy, we hear the beat of the solar drums.

Absence makes the heart grow fonder, perhaps. Solar energy has become the romanticized solution to everything — energy, pollution and safety. It's and Mother's Milk and Apple Pie. The news media wax eloquent about solar energy every time a local solar enthusiast insulates his house and puts up a showpiece solar collector.

In this chapter we will hear from many high-profile headline-makers, such as Ralph Nader. Another is Denis Hayes, Jimmy Carter's politically appointed director of the Solar Energy Research Institute (now the National Renewable Energy Laboratory), and head honcho of Earth Day 2000. Christopher Flavin is the president of the Worldwatch Institute; Lester Brown is its founder. Michael Oppenheimer is head of the Environmental Defense Fund. Amory Lovins founded the Rocky Mountain Institute, a group that promotes energy conservation. Russell Train is Chairman of World Wildlife Fund and The Conservation Foundation. He was the first chairman of the President's Council on Environmental Quality and was director of the Environmental Protection Agency. Senator Gaylord Nelson (D. Wisc.) is founder, along with Denis Hayes, of Earth Day. There are many solar cheerleaders, typically celebrities, who lend their backing to solar causes.

In some cases, solar enthusiasts merely reflect a rosy — but misinformed — view of the future. Some solar enthusiasts set themselves up as moral authority. In other cases, their comments barely disguise a coercive or manipulative intent. And there is a steady barrage of statistical obfuscation. Next come the conspiracy mongers who tell us that solar energy is being suppressed by (who else?) Big Oil.

Solar Cheerleaders

Surprisingly, the Hollywood crowd that shows up at pro-solar rallies produces no documentation, save the news reports saying that they were there. For example, at Earth Day 2000, Chevy Chase, James Taylor, Clint Black, Tom Arnold, Ed Begley Jr., Robert Kennedy Jr., Gene Karpinsky and Winona LaDuke joined Leonardo DiCaprio, Al Gore, and Denis Hayes in endorsing a generic statement calling for "for clean power, clean air, clean cars and clean investments," obviously railing against everybody who prefers dirty power, dirty air, and dirty investments. Mostly, they expressed themselves by chanting "Clean Energy Now!" egged on by actor Esai Morales, but left no written record of how their vision was to be accomplished.

In November, 2000, British Petroleum honored Maya Angelou, Richard Dreyfuss, Jane Fonda, Harrison Ford, Mikhail Gorbachev, Vice President Al Gore, Quincy Jones, Kermit the Frog, Joni Mitchell, Randy Newman, Robert Redford, Diane Sawyer, Sting and Ted Turner for "their commitment to the environment and to showing the world the value of solar energy." Their pro-solar sympathies, along with those of Robert Redford, Barbra Streisand, Ed Asner, and Danny Glover are well known. Unfortunately, it is very hard to find any written details that would be useful for implementing their goals. Al Gore's book, *Earth in the Balance*, is short on science and long on a legislative agenda.

Pollyannas

Part of the solar drumbeat comes from Pollyannas who have an ecstatic devotion to the Sun. They show continual optimism despite repeated failure of solar predictions to come true.

> "Everything will be solar in 30 years."
>
> Ralph Nader (1978)

> "… 50% solar by the end of the century"
>
> Denis Hayes (1978)

> "Solar. Do you want me to give you a book that shows how realistic solar is? We've got windpower, biomass, photovoltaics, tidal — all kinds of technology."
>
> Ralph Nader (1997)

> "In the coming decade, solar energy use in the U.S. is expected to grow dramatically, eventually producing enough thermal and electric clean energy to supply a million homes …"
>
> Carol Browner (1998)

"The California Energy Commission projects a renewable share in the state's electricity of 50 percent to 60 percent by the year 2000."

John O. Blackburn (1987)

"Within a few decades, for example, the United States might get 30 percent of its electricity from sunshine, 20 percent from hydropower, 10 percent from biomass, 10 percent from geothermal energy, and 10 percent from natural-gas-fired cogeneration.

Brown, Flavin & Postel (1991)

"Optimists believe that ten years will pass before photovoltaics can seriously compete with fossil-fuel electricity sources at a reasonable price."

Oppenheimer & Boyle (1990)

"Our conclusion is that with a strong national commitment to accelerated solar development and use, it should be possible to derive a quarter of U.S. energy from solar by the year 2000. For the year 2020 and beyond, it is now possible to speak hopefully, and unblushingly, of the United States becoming a solar society."

(Carter) White House Council on Environmental Quality (See Reese, 1979)

"Photovoltaics has the advantage of a relatively short time period (1 to 2 years) required to bring a large (1-GW[$_e$]) power plant on line ... researchers at Chronar Corporation have optimistically estimated ... approximately 10 GW[$_e$] per year of PV manufacturing capability could be in place by 1995, leading to *40-GW[$_e$] installed capacity by 2000*." [emphasis added. In reality, the US had 275 MW$_e$ — 0.275 GW$_e$, not 40 GW$_e$, as of 1998.]

H. M. Hubbard (Former head of SERI) (1989)

It is now 2001. Nader's prediction for 2008 will certainly be no better than the rest of them were for 2000. Scientists and engineers simply did not — and do not — make such lofty predictions.

> "Developing countries have the opportunity to move straight into the solar age...
>
> "Many find it surprising that renewables—primarily biomass and hydropower—already supply about 20 percent of the world's energy."
>
> Brown, Flavin & Postel (1991)

The intrepid scholars from the Worldwatch Institute have it backwards. More logically, they should say that the supply of energy from renewables has already *dropped* to 20 percent. (It used to be 100%.)

But the 20% figure is probably an overestimate for 1991 anyway. According to the Energy Information Agency,[16] "The renewable share of total world energy consumption is expected to decline slightly, from *9 percent in 1999* to 8 percent in 2020, despite a projected 53 percent increase in consumption of hydroelectricity and other renewable resources." [emphasis added]. That nine-percent figure may be an underestimate for 1991, but if the EIA expects the solar contribution to drop by one percentage point in the next 20 years, the drop from 1991 to 1999 was unlikely to have been 11 percentage points.

> By the year 2000, such renewable energy sources could provide 40 percent of the global energy budget; by 2025, humanity could obtain 75 percent of its energy from solar resources ...
>
> Denis Hayes (1977)

Obviously, Hayes's prediction for year 2000 was far-fetched, though the obvious failure did not keep him from leading the troops at Earth Day 2000 (an event that was overshadowed by a six-year-old Cuban boy named Elián Gonzales). In 1999, about nine percent of the energy used around the globe came from solar sources, not appreciably different from what it was when Hayes made this erroneous prediction. The solar contribution did *not* double in the two decades 1977–1998, let alone become 40% of the energy budget in 2000.

One doesn't need the benefit of hindsight to spot the errors in such flamboyant predictions. Firewood is a very limited resource everywhere that it is regularly used as the primary energy source (such as equatorial Africa) so *that* contribution won't double. Most of the largest, easiest-to-use hydropower sites (such as Niagara Falls and Hoover Dam) everywhere

---

[16] *International Energy Outlook 2001,* Energy Information Agency (DOE/EIA-0484, March 2001).

in the world are already in use. To double the output of hydropower would obviously involve tens of thousands of small dams. To double the solar contribution would require developing wind, photovoltaics, and solar heat so much that they would rival wood and large hydro in a mere ten years. No energy source in history has ever taken over that much of the market in less than forty years. All that have made significant inroads have been strong sources — first coal, then petroleum, then natural gas, then nuclear.

The Union of Concerned Scientists weighed in with a timetable (Table 1) for solar energy to take over. Table entries are in quadrillion BTU (quads, an archaic unit), except for the bottom row, which is expressed in percent of total energy. Notice that the US energy consumption was supposed to drop to 82.7 quads by 2000; in actuality it increased to about 97 quads. Aside from those for hydropower, all predictions in the table are grossly in error.

Table 1: The prediction of the Union of Concerned Scientists (1991) for solar energy to take over the US energy supply. (From Brower, 1992)

	1990 Actual	2000 Predicted	2030 Predicted
Biomass	3.45	7.65	18.38
Hydropower	3.31	3.61	3.77
Geothermal	0.18	0.57	3.79
Solar	0.08	0.60	4.59
Wind	0.04	0.68	3.63
Renewable Total	7.06	13.10	34.52
U.S. Consumption	89.0	82.7	65.3
Renewable Fraction	7.9%	15.8%	52.9%

"Solar energy may well become the primary energy source for America and the world in the 21st century."

Senator Edward M. Kennedy (see Berman & O'Connor, 1996)

Platitudes, platitudes. Worldwatch is not to be outdone either.

"Norway, for example, relies on hydropower and wood for more than half its energy. ... In Israel and Jordan, rooftop solar collectors already provide 25-65 percent of domestic hot

> water. ... Sub-Saharan Africa derives three quarters of its
> energy from wood ..."
>
> Brown, Flavin & Postel (1991)

As teenagers are wont to say these days, *Well, DUH!* Word has it that
the Israelis have consulted with the Norwegians on how to develop their
Sinai Fjords.   In turn, the Norwegians are studying rooftop solar water
heaters.

Additionally, these world-watchers have failed to notice that deriving
three-quarters of the energy from wood in sub-Saharan Africa has had
devastating environmental effects.

> Very often fuelwood use is unsustainable, accompanied
> by deforestation and all the attendant environmental ills. ...
>
> Paul & Anne Ehrlich (1991)

The Pollyanna group has its Let-them-eat-cake contingent.   We start
with the Ehrlichs, who might like everybody to move to waterfronts.

> One of the cheapest forms of transport (in terms of energy
> consumption) is by water — barge and riverboat.
>
> Paul & Anne Ehrlich (1991)

> "Soft technologies ... capitalize on poor countries' most
> abundant resources (including such protein-poor plants as
> cassava, eminently suited to making fuel alcohols)..."
>
> Amory Lovins (1977)

> "John Sheffield from Oak Ridge National Laboratory
> postulated that with the likely depletion of most fossil fuels
> by 2100, alternate energy resources will be developed
> according to the region's indigenous resources..."
>
> "Energy Alternatives Vital to Meet Future
> Demands"
>
> *American Physical Society News*,
> July 1996

> "At length I recollected the thoughtless saying of a great
> princess, who, on being informed that the country people had
> no bread, replied, 'Let them eat cake."
>
> John Jacques Rousseau
> *Confessions* [1781–1788]

Rousseau recognized that the princess was thoughtless. The Ehrlichs, Amory Lovins and *APS News* are not so perspicacious.

Let the developing countries use the sunbeams and chicken manure they've always had. Anything less would be uncivilized.

High Priests

But for the topic, there is little to distinguish the high priests of energy telling us it is sinful to use energy from the Taliban telling the beleaguered Afghanis that it is sinful for women to show their faces in public.

Citing the wisdom of Amory Lovins, Berman and O'Connor tell us that using energy is morally wrong.

> [Amory Lovins's wisdom]: "For over 90 percent of energy uses, electricity is an indefensible luxury."
>
> Berman & O'Connor (1996)

Lovins himself makes it clear that he is opposed to *any* source of abundant energy:

> "… if nuclear power were clean, safe, economic, assured of ample fuel, and socially benign per se, it would still be unattractive because of the political implications of the kind of energy economy it would lock us into."
>
> Amory Lovins (1977)

> "We can and should seize upon the energy crisis as a good excuse and great opportunity for making some very fundamental changes that we should be making anyhow for other reasons.'"
>
> Russell Train (1974)

> "Even so, fossil fuels will be difficult to abandon because they have become an addiction."
>
> Oppenheimer & Boyle (1990)

Face it. We don't need food; it's an addiction. We don't need housing; it's an addiction. We don't need clean water; it's an addiction. We don't need refrigeration; it's an addiction. We don't need electricity; it's an addiction. We don't need transportation; it's an addiction.

High-Priest Russell Train makes no secret of his view of the moral superiority of "environmental and social policy objectives," nor of his intent to manipulate the public to achieve them.

> "Regulators should tap the power of the marketplace to provide the economic signals and financial incentives *necessary to achieve environmental and social policy objectives.*" [emphasis added.]

Russell Train (1993)

> Over the next decade or two, Americans should try to cut their per capita energy consumption *in half.* [emphasis in original] ... Nothing less than a reorganization of the American way of life is *required* [emphasis added] .

Paul & Anne Ehrlich (1974)

Sieg Heil!

> "...one is forced to ask the churches, 'Where have you been on this [environmental] issue?'"

Russell Train (1990)

Puppeteers

Cheerleaders and Pollyannas are one thing; puppeteers are yet another. Their incessant drumbeat tells us that the only obstacles to solar energy are political, and they seek to coerce and manipulate people by any and all means available to adopt solar energy. Some prefer carrots. Others prefer sticks. All are manipulative.

Here are several examples.

> "Simply by creating the incentive of higher energy prices, perhaps through a higher gasoline tax, the government could encourage considerable movement toward efficiency..."

Paul & Anne Ehrlich (1991)

> "Pacific Gas and Electric, Southern California Gas Southern California Edison and San Diego Gas and Electric should be required to undertake demonstration financing programs for solar water heaters. The utilities should be given substantial flexibility within certain parameters, to develop these programs."

California Public Utilities Commission spending
the utilities' money (1980)

"It's a matter of changing the pattern of practice, getting people who can't afford the upfront cost over the hurdles by giving them a tax credit to make it an offer that they can't refuse, and then change will take place."

Al Gore (2000), making his $48 billion "offer that they can't refuse" to manipulate the public to use solar energy (See tidalelectric web site)

"Some of the most astute political thinkers among the solar activists, including Earth Day organizer Denis Hayes, realized that solar would die on the vine without high-level *political*, technological and *financial support...*" [emphasis added]

"Meanwhile, R&D support for solar water heating was almost impossible to secure.

Berman & O'Connor (1996)

Berman & O'Connor shed crocodile tears over the loss of government support for solar water heating. Designing a domestic water pre-heater is not rocket science. Any competent carpenter can team up with any competent plumber to build one. In fact, a good tradesman of either profession could do the job without the help of the other.

"But coercion is not necessary ... Why use penal legislation to encourage roof insulation when tax incentives and education (leading to sophisticated public understanding now being achieved in Canada and parts of Europe) will do?"

Amory Lovins (1977)

It's not as if the American Wind Energy Association (www.awea.com) has a vested interest, but — perhaps coincidentally — they do seem to favor subsidies for wind turbines.

[headline] "Minnesota State Panel *Orders* Utility to Install 400 MW[e] of Additional Wind Capacity" [text] "22 January 1999: Minnesota Public Utilities Commission, in a 4-0 vote *ordered* Northern States Power Co., a Minneapolis-based investor-owned utility, to acquire another 400 megawatts of wind generation by the year 2012. [emphasis added]

http://www.awea.org/news/index.html

> [headline] "Wind Group Applauds New York Commitment to Renewable Energy" [text] "13 June 2001 - Gov. Pataki *issued an Executive Order on June 11th requiring all agencies* of New York State, including the New York City Metropolitan Transportation Authority, to purchase 10% of their electricity from renewable energy sources by 2005 and 20% by 2010." [emphasis added]
>
> http://www.awea.org/news/index.html

Denmark has decided [2001] that subsidies are no longer necessary to coerce the conversion to solar energy. All they have to do is to levy confiscatory taxes on all other energy supplies. Simple!

Berman and O'Connor berate Bill Clinton for not being able to coerce a little more tax money from US citizens:

> "Despite Clinton's expression of concern in January 1993 about 'carbon taxes' and global warming, he couldn't even pass a 4.3-cent-per-gallon gasoline tax increase because he couldn't even persuade his own party to support stronger measures."
>
> Berman & O'Connor (1996)

> "However, whether or not this age-old friend of man [*i.e.,* solar energy] will be given a share of our future energy markets will depend (and this seems somewhat profane) upon its economic viability in competition with the prosaic physical systems devised by man."
>
> Arthur R. Tamplin (in Commoner, 1975)

And why is it profane that solar energy is required to compete on its own merits *vis-a-vis* other energy sources? And what, pray, is prosaic about the fantastic machines that serve us every day? (Five will get you ten that Tamplin does not make his own toothbrushes.)

Clinton appointees Federico Peña (former DOE head) and Carol Browner (head of the Environmental Protection Agency) added thunder from their Washington offices.

> "To increase the availability and affordability of this and other renewable energy sources, we should invest in such technologies as advanced wind turbines, co-firing of coal plants with forest and agricultural biomass and on-site solar and photovoltaics to enable buildings to become net energy

producers. The Million Solar Roofs initiative provides a goal within our reach."

Federico Peña (1998)

Perhaps that (million-roof) goal is within reach and, if so, they'll be able to do it with just tax money. Moreover, during the time the homes are being decorated with solar collectors, more than a million new homes will have been built *without* solar collectors.

The Environmental Protection Agency is not under Congressional mandate to subsidize solar energy projects, but such restrictions did not restrain Clinton appointee Carol Browner from using tax money for unintended purposes.

> "To help accelerate the diffusion of solar energy into the marketplace, EPA has initiated a number of solar-related activities.
>
> "EPA and the utility industry are partners in a $4.25 million cost-shared program to fund the installation of 30 commercial and residential rooftop photovoltaic systems totaling 376 kilowatts of generating capacity."

Carol Browner (1998)

In passing, we note that 376 kW_e is about a millionth of the electrical power consumption of the US. Moreover, Browner is talking about *capacity*, which is 5 or 6 times as large as average generating power. The program was a government give-away combined with an extortion of utilities' money to pay for a feel-good project that would do nothing of any value.

As long as we're into centrally controlled economics, note that *gasoline* could be made free if the government would simply buy it for us. The US government and California together offered 95% tax credits for California's first wind farms. Al Gore had more ideas for giving away tax money to promote his anti-energy agenda:

> "Consumers would get tax credits of up to $6,000 for buying electric cars and $2,000 for energy-efficient new homes under a campaign proposal by Vice President Al Gore."

News-Journal Wire Services (2000)

Everybody jumps on the solar energy bandwagon. Carol Browner gave credit to the EPA for preventing pollution through use of renewable technologies (but complimented them only for the five years the Clinton administration had then been in office).

> "Over the past five years, we have worked to find new common-sense ways to prevent pollution, which is why our Agency enthusiastically supports solar and other renewable energy technologies."

Carol Browner (1998)

In the comment below, Michael Brower tells us that there is no need to fix the roof when the sun is shining.

> "Electric utilities maintain a reserve capacity (typically 20 percent in excess of peak demand) to allow for plant shutdowns. This reserve should suffice until solar and wind energy constitute at least a few percent, and possibly as much as 20 percent, of the total electricity supply ..."

Michael Brower (1992)

In other words, "Stop building power plants. Follow my plan." It seems that California did just that.

Or, maybe we should start stoning sinners.

> ... vacationing by automobile should be discouraged ...
> Three-day weekends ... can be abolished ...

Paul & Anne Ehrlich (1974)

Politicians like to talk about creating jobs. Let's remember that long ago, the *only* job was seeking food. Nowadays we do not have to be hunters and gatherers, because all of our food is produced by about 5 percent of the population.

Abundant energy creates jobs, but most of them are not in the production and transmission of energy. The jobs are in industries that *use* energy.

In colonial New England, a major occupation was that of cutting trees for firewood. Now our energy from coal, natural gas, uranium, and petroleum, whether as electricity or as fuel, is so abundant and easy to use that very few of us even know anybody directly whose job it is to bring us the energy.

But some people think that the act of producing energy should be labor-intensive. For example, Lester Brown *et al.* [Brown, Flavin & Postel (1991)] tell us that nuclear power provides 100 jobs per thousand gigawatt-hours a year. In those units, geothermal power employs 112 and coal 116. More to their liking is solar-thermal, which employs 248, and wind, which employs 542. By their reckoning, nuclear is worst and wind is best. Wind keeps the most people employed bringing us energy. Brown *et al.* are not alone.

> "The total number of jobs in the U.S. increased by 41 percent between 1950 and 1971, while the total number of jobs in the energy producing industries increased by 5.5 percent, despite a doubling of energy production."
>
> Berman & O'Connor (1996)

> "In fact, every quintillion joules/year of primary energy fed into new power stations *loses* the U.S. economy some 71,000 net jobs, because power stations produce fewer jobs per dollar, directly or indirectly, than virtually any other major investment in the whole economy."
>
> Amory Lovins (1977)

The annual amount of energy fed into power stations has increased by 7.5 quintillion joules since 1973, and the civilian labor force has increased by some 20 million.

> "A lot of the money goes into areas that are not job producing and don't help the economy. Nuclear energy is one example…"
>
> Jane Fonda (teamhouse web site 2000)

Fonda misses the point. Whereas mechanization improves efficiency and eliminates some jobs, abundant energy *allows* jobs to exist that otherwise would not. *All* energy produces jobs and *all* energy helps the economy.

Statistical Smoke Generators

Darrell Huff's classic book, **Lying with Statistics**, is a beautiful study, with examples taken from popular and professional literature, of the misuse of statistics. Here are similar examples taken from solar-energy literature.

> … Use of gas, coal and oil rose by 4.5, 2.3, and 1.8 percent respectively, reaching all-time highs. At the same time, geothermal energy increased by 5.5 percent, solar power by 16 percent, and wind energy by 26 percent."
>
> *Chicago Tribune* (May 25, 1997)

I gave a friend one penny last year, and two pennies this year. That was his biggest increase (100%!!!) in income. He got a trifling 3% raise ($2400) in his ($80,000) salary. Obviously, one-hundred percent is better. We hear that Maude's Refreshing Hippopotamus-Tooth Tonic is about to steal market share from Coca-Cola. Production rose 94% last year alone! (We return to this topic in Fig. 17.)

> "Since 1990, wind power has risen 150 percent, representing an annual growth rate of 20 percent,…."
>
> Worldwatch (1996)

If a quantity grew at an annual rate of 20% for six years, the quantity would *triple* in that time, corresponding to an increase of 200 percent.

> "The market prefers other options. In the 1990s, global nuclear capacity rose by 1% a year, compared with 17% for solar cells (24% last year) and 24% for wind power--which has lately added about 5000 megawatts a year worldwide, as compared with the 3100 new megawatts nuclear power averaged annually in the 1990s."
>
> Lovins & Lovins (2001)

This statistical smokescreen is more insidious than that of the *Chicago Tribune* because it protects the truth by a bodyguard of lies, the very least of which is the emphasis on large percentage growths of trivial numbers. Here is a sample of what is wrong with the quote from the Lovinses.

➤ Total worldwide electric capacity from solar cells is less than that of one single typical nuclear power plant. Moreover, the solar cells produce power 15 to 20 percent of the time; nuclear power plants produce power 80% of the time, and recently in the US, 90% of the time.

➤ The increase in delivered energy from nuclear power plants increased from 1722.5 billion kWh in 1990 to 2416.4 billion kWh in 1998. This amounts to an annual increase of 4.2% per year, *not* 1% per year.

But it's not even that simple. A 1% per year increase compounded over eight years would amount to an overall increase of 8%. The actual increase over only eight years (1990–1998) was 40%, corresponding to a compound annual increase of 4.2% not the 1% claimed by Lovins. When Lovins says, "global nuclear capacity rose by 1% a year," it gives an entirely different impression to most people than if he had said (honestly) that global nuclear energy production rose 40% in only eight years. See the actual energy production in Fig. 17.

➤ The 8-year average increase in nuclear energy production amounted to an around-the-clock average power increase of 9900 MWe. Assuming a capacity factor of 80%, this is an increase of 12,300 MWe capacity in eight years, not the 3100 MWe claimed by the Lovinses for ten years.

➤ The most common statistical lie in the solar-energy game is to report the *capacity* as if it were the same as the production. The 5000 MWe figure is the growth in wind *capacity*. The around-the-clock average power output would be about 30% of that (if they're lucky), or about 1500 MWe of the annual growth in wind power *production*. This is about 15% of the average annual growth in production from nuclear power plants. Only a polished obfuscator could make an annual increase of 9900 MWe in steady nuclear power look puny compared to 1500 MWe of stochastic wind power.

The Lovinses are not alone in this deception. Flavin and Dunn of Worldwatch has produced a graph[17] showing that nuclear power grew at the insipid rate of 0.6% from 1900-1997, which would correspond to a 4% increase during the period. Their graph was also printed unquestioningly by *The Economist*[18]. The Worldwatch data are given in Table 2. Notice the emphasis on *percentage* growth and the values of 0.6% for the annual growth of nuclear power. (This would result in a total of 6% growth over 10 years; in fact, nuclear grew by 40% in only 8 years.)

---

[17] http://www.jxj.com/magsandj/rew/1999_04/comingofage.html

[18] "Special Report, Nuclear Power: A renaissance that may not come," *The Economist* (May 19-25, 2001)

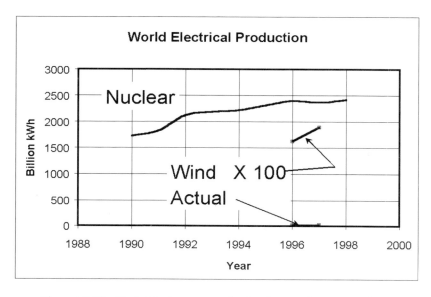

Figure 17: World electrical energy production from nuclear power plants and from wind (1996 and 1997, assuming a generous 30% capacity factor). The contribution from wind has been exaggerated by a factor of 100 so that it would show on the graph.

Table 2: Global Trends in Energy Use, 1990-2000 According to Worldwatch

Source	Average annual growth rate* (percent)
Wind power	25.1
Solar photovoltaics	20.1
Natural gas	1.6
Oil	1.2
Nuclear power	0.6
Coal	−1.0

* "Based on installed capacity for wind and nuclear power, shipments for solar PV, and consumption for natural gas, oil, and coal."

Bill Clinton developed a plan to use taxpayers' money for a feel-good, do-nothing solar project (see Fig. 18). This case is such an egregious example of lying with statistics that it deserves special mention. The chart in Fig. 1 was trotted out to impress the audience.

"Speaking at the United Nations Thursday, Clinton called for photovoltaic panels to be installed on rooftops as part of his plan to reduce dependence on fossil fuels that are believed to cause global warming. 'By capturing the warmth, we can help turn down the Earth's temperature,' he said."

Boston Globe (**June 28,** 1997)

Let us have a hard look at the data. Carbon dioxide-induced global warming will raise the temperature of the Earth somewhere between zero and 5 °C during the next century. Choose a number. Any number. Call it ΔT, ("delta-T"), the increase in temperature.

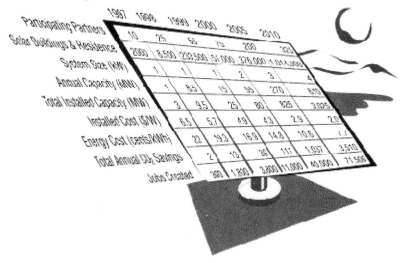

Figure 18: The marvelous effects of Clinton's Million Solar Roofs Initiative. By 2010, there should be 3,025 MWe of installed PV capacity (500 MWe average), and a CO_2 savings said to be 3,510 thousand tons of CO_2.[19]

---

[19] http://www.es.wapa.gov/pubs/esb/97Oct/at_roof.htm

The amount of global warming will depend on the amount of CO_2 humanity puts into the air. The US fraction of the CO_2 put into the air is about 20%. Therefore, the US is responsible for 20% of ΔT, whatever ΔT might be. If your choice was 5 °C, then the US would be responsible for 1 °C of that temperature rise.

Million Solar Roofs Program
Reduces Global Warming!

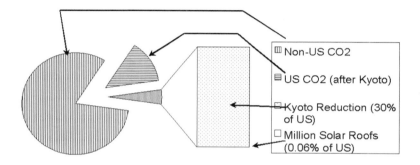

Figure 2: CO_2 production in the world causes global warming. The non-US fraction is about 80%. The Kyoto Protocol would force the US to decrease its production by 30% of the expected 2010 production. The Million Solar Roofs Initiative would be responsible for a reduction of 0.001%.

According to the Kyoto Protocol, by 2010, the US is required to reduce its CO_2 production to 7% below 1990 values, amounting to 30% of the otherwise expected 2010 CO_2 production rate. Therefore, Kyoto would be responsible for reducing ΔT by 6% (*i.e.,* 30% of 20%). If your choice of ΔT was 5 °C, then the Kyoto accords would reduce the US's contribution by 0.3 degrees. Instead of a ΔT being 5 °C, it would be 4.7 °C.[20] (See Fig. 2.) If

---

[20] The "Kyoto accords" (*not* treaty) exempt the vast majority of the world's population (including India and China) from CO_2 restrictions. European heads of state have been all over George W. Bush because he has failed to

your choice for ΔT was 0.5 °C (just as that of the last century was), then the Kyoto accords would limit the US's contribution to 0.03 °C.

Despite broadcasting a barrage of global-warming scares over the years, National Public Radio and ABC News (7/19/01) have actually recognized that the Kyoto Protocol would have very little effect on global warming. However, they seem to want that screen door on the submarine anyway, just for decoration.

Trifling, you say! Indeed, but what about solar roofs? The amount of CO_2 reduction [see Fig. 1] to occur as a result of the Million Solar Roofs Initiative is supposed to be 3.5 million tons, a mere 0.06 percent of the 6000 million-ton US CO_2 production (Fig. 2). Therefore, we can reduce the earth's temperature by 0.0029 °C if you chose the extreme 5 °C for ΔT and 0.000 29 °C if you chose 0.5 °C for ΔT.

> "'By capturing the warmth, we can help turn down the Earth's temperature,' he [Bill Clinton] said."
>
> Boston Globe (1997)

But what do you expect from households with only 1,014,000 solar collectors purchased at taxpayers' expense? Their next project will be to make the earth rotate a little slower by all driving our cars east.

The most common statistical trick solar advocates use is to tell us the *capacity* of a proposed solar project, not the year-round expected output. Almost invariably, the year-round average power output of a solar project is about 15 to 20% of the capacity. (Exception: Modern wind turbines can produce about 35% of their capacity, as we shall discuss in Chapter 9 and in Appendix B.)

> "The construction costs for U.S. nuclear-power plants in the mid-1980s averaged $2500 per kilowatt, *exactly the same cost* that the Chronar Corporation <u>anticipates meeting</u> with its planned photovoltaic plant." [underline added]
>
> Oppenheimer & Boyle (1990)

Three problems. "Anticipates" is one thing, reality is another. The anticipated cost reductions did not occur. More importantly, Chronar is talking about *peak* kW_e. A nuclear power plant (capacity factor = 80%) will produce about five times as much energy in a year as a solar plant (capacity factor = 16%) even if one were huge enough to have the same nameplate power rating. Finally, the "costs" of photovoltaics in Oppenheimer's

---

endorse the Kyoto accords. Never mind that they haven't done so either, nor that the Senate has already rejected it 95 to zero.

discussion include only the costs of the photocells themselves, not the cost of a complete power plant.

> "Meanwhile the nuclear option will linger on, not alive but not quite dead, beset by continuing failures, the lack of permanent waste disposal, the potential for proliferation of weapons-grade fuel, and excessive costs."

> Oppenheimer & Boyle (1990)

Fast-forward to 2001: There are 438 nuclear power stations around the world with a capacity of 351,000 MWe. There are 31 under construction for another 29,900 MWe.

All nuclear waste at all US power plants that have been producing power since 1957 is still stored at the power plants. Try storing decades' worth of wood ashes, coal ashes, or carbon dioxide! Oppenheimer has not yet been able to figure out the obvious: there isn't much waste.

Another standard pro-solar statistical lie is hidden in capricious grouping. For example:

> "About one-fifth of all energy used around the world now comes from solar resources: wind power, water power, biomass, and direct sunlight."

> Denis Hayes (1977)

Sure. And well over 95% of the milk sold in the US comes from cows and bears. There are two sources that supply the lion's share of the solar energy, biomass and waterpower — the traditional sources. The two piddle-power sources — wind and direct sunlight — provide the bears'-milk fraction of the total. (Note again the incorrect figure. In 1977, the world got 9%, possibly 10% in 1977 — not one-fifth — of its energy from renewable resources.)

The Don Quixotes

To Don Quixote, a windmill was an enemy to be lanced. To energy-conspiracy theorists, windmills are friends and Big Oil is the enemy. The only difference between Don Quixote and the conspiracy theorists is what side they're on.

The starting point for conspiracy theorists is that solar energy is a matter of politics, not science and engineering. The drum they beat says that if you gave Exxon a solar depletion allowance, we'd be running on solar energy tomorrow. Enjoy these comments.

"Stealing the sun is too ridiculous to imagine — or is it?

"Monopolization of the sun. A virtual enslavement of generations to come ... control of a man's life through the control of the energy necessary to his existence.

"Giant corporations are setting out to ... own the sun outright. The solar market equals 71 times Exxon's annual sales."

John Keyes, *The Solar Conspiracy* (1975)

"The only reason why solar energy has not yet been developed is that the oil companies don't own the sun."

Ralph Nader (1978)

"The suspicion is almost unavoidable that giant firms, because of their large investments in nuclear technology, hope that solar energy will not gain rapidly"

Senator Gaylord Nelson of Wisconsin (in Keyes, 1975)

"[Energy barons] had to kill enthusiasm for solar..."

Berman & O'Connor (1996)

"Here's the documented proof [the Berman-O'Connor book] that America's utilities are crippling—and attempting to eventually own—solar energy. Read this book. The alternative is another century of energy slavery."

Richard and Karen Perez,
Publishers of *Home Power*
(Berman-O'Connor Book Jacket, 1996).

> "This well-documented, passionate book by Berman and O'Connor is a jolting reminder of how ... already-practical solar energy was marginalized by the giant fossil fuel and nuclear power corporations."

> "[Solar energy] is not owned by Exxon or Peabody Coal. But its availability for heating, lighting, and cooling America can be controlled by these and other non-renewable fuel corporations."

Ralph Nader (Foreword to Berman/O'Connor, 1996)

The main thrust of the Berman-O'Connor book is that energy barons in big corporations are successfully keeping solar technology from the public. Those villains are obviously paying bribes to all of the plumbers and carpenters in the US to keep them from building solar water heaters and the like.

> "Every essential feature of the proposed solar transition has already proven technically viable; if the 50-year timetable is not met, the roadblocks will have been political — not technical."

Denis Hayes (1977)

Hayes is speaking of *his* timetable, of course.

> "NADER: But we're not even close to developing a national program for converting to solar energy."

> "PLAYBOY: Why not?"

> "NADER: Because Exxon doesn't own the sun."

Playboy, June, 1992.

If I chide these solar enthusiasts too harshly for your taste, please bear this in mind: I am, with apologies to none, *pro-energy*. I am in favor of solar energy wherever and whenever it is practical. I have no patience with delirious Pollyannas, societal manipulators, statistical obfuscators, or conspiracy theorists.

The rest of the book will go into details of the various manifestations of solar energy and ask how well they will be able to meet the demand for energy.

Chapter 4. Questions for Inquiring Minds

Fortunately, not everybody is bedazzled by solar Pollyannas and manipulated statistics. In this chapter, we consider questions that must occur to inquiring minds.

For one, everybody realizes that man has inhabited the earth for a long time, and has used solar energy for millennia. If sunlight is abundant and easy to use, one may ask, why doesn't everybody simply use it, putting Exxon and the utilities out of business? Why *has* it diminished in importance from 100% down to about 7% in the US?

Questions about Home Heating

There is an abundance of books, newspaper articles, and magazine articles about using solar energy to heat homes. The style of writing varies from one to the other, but the rules for accomplishing the task do not. The rules immediately induce questions about solar energy generally.

The most important rule is

Rule #1: Insulate your house so well that *even* solar energy can heat it.

The rule is rarely stated this way, but *every* article on the subject of solar heating emphasizes how well a house must be insulated. Authors usually sugarcoat the treatment, telling you how morally correct you will be for doing so, and (correctly) how much more comfortable the house will be, but the message is always clear. You *must* insulate your house extremely well if you intend to use solar energy to heat the house.

"Energy Efficiency: First Things First"

"So the first step is to make your home and appliances as efficient as you can."

Amory Lovins (2001)

"The solar path is simple: use less energy."

Berman & O'Connor (1996)

Would that advice be offered if solar energy were a strong source, such as a nuke in the basement?

Rule #2: Install a heat collector as large as the roof.

The best candidates for solar-heated homes are those with large roof areas that face south. No expert says, "put in enough collectors to cover 10% of the roof." You have to cover *most* of the (south-facing) roof; the more the better.

Again, would that be necessary if solar energy were a strong source of heat?

A friend had a modest house built about 20 years ago and had a solar heat collector installed. Being young and naïve, he followed the (non-expert) builder's advice and had only one collector installed, which covers only about 20% of the south-facing part of his roof. Later, he criticized the builder for selling him only one, because — nice as it is — it simply doesn't provide enough heat.

Still, despite his desire to have more solar heat, the friend has not installed more collectors. Why is his professor's salary inadequate for him to afford a retrofit?

Rule #3: Install a *real* heating system as a backup.

Because the sun does not always shine, and because the weather may be too cold, it is always necessary to have a backup system for the solar system.

Would not *all* solar systems — for heat or otherwise — similarly require backup systems?

Rule #4: Install a storage system if you intend to have solar energy supply most of the heat.

A solar system obviously cannot supply heat when the sun is not shining — well more than half of the time in winter — or when the sun's rays are not directly falling on the collection system. There must be a storage/retrieval system if plans call for any solar heat to be delivered at such times.

Would not all solar systems require storage systems?

Would not all solar systems have to be overbuilt to "stock up" while the sun was shining?

How could you store high-temperature heat that might be needed for smelting iron, for example?

Questions about the Impact

Often, people speak of solar energy only in reference to home heating. Indeed, that is one of the best uses for solar energy. But society uses energy for other purposes as well. Suppose we were to pull out our magic wands and use solar energy for *all* of our space heat and hot water, both for homes and commercial establishments. How would that affect the total energy picture?

The inquisitive individual would know that domestic space and water heating would only be part of the total energy picture. Where would the energy come from for transportation, manufacturing, farming, refrigeration, air conditioning, and pumping the water that gets heated by the domestic solar water heaters? What about electricity for lighting, communications, and medical equipment?

Questions about Nighttime and Weather

Society uses energy around the clock, all seasons of the year, in good weather and in bad. It must therefore occur to rational people to ask: How do we use solar energy when the sun is not shining? Is *this* the reason we don't use it?

For only the cost of a lot of solar collectors and the necessary wiring, you can read by bright electric lights when the sun is shining. To run the same lights at night involves a few additions to the simple solar system. An inquisitive person might wonder how solar energy can be used when the sun is not shining, either because it is night or because the weather is overcast.

> "To put it another way, it would cost five dollars a month to light a hundred-watt lamp for six hours a day with photovoltaic energy, as opposed to two dollars a day with standard power."
>
> Oppenheimer & Boyle (1990)

How about six hours a *night?*

The inquisitive reader must wonder how large a bank of backup batteries would have to be to supply backup power for such energy intensive devices as clothes dryers. How rapidly can they be recharged? How big must the solar-cell array be to gather enough energy during a short day of winter sunlight to charge batteries enough to last for many cloudy days and nights? How much do the batteries cost, and how many must be used? How long might they last before they need to be replaced? What is the environmental impact of battery disposal/recycling?

Of course, a solar system can be built purely as a supplemental system without a backup *solar* system. But there is a backup system nevertheless, namely, the energy distribution we already have in place.

> "No storage batteries are required because each house has two electric meters, one to measure electricity consumed (when the PV modules are not generating enough power) the other to record the amount *sold* to the utility (when the PV panels produce more than is needed).
>
> Oppenheimer & Boyle (1990)

The astute reader is entitled to wonder whether Oppenheimer and Boyle expect to cure our "addiction" to fossil fuels by relying so heavily on utility power.

Here is an admission that photovoltaics can never amount to much. Oppenheimer expects photovoltaics to provide only supplemental energy, *i.e.,* less than 20% of the electricity under the very best of circumstances. The utility would not be able to get rid of any of its generating capacity, because it would frequently be required to provide 100% of the power.

Questions about US and International Politics

A reasonable question, then, is this. During night or inclement weather, why not receive energy from another place in the world where the sun is shining? For example, when it is nighttime in both New York City and the Sonoran Desert in Arizona, how could we get energy from (say) the Saudi Arabian Desert?

And are we sure that we want to go that route? Wasn't there something about dependence upon the energy sources in the Mid East?

Questions about Seasonal Variations

The problems of winter must surely be triply, possibly quadruply bad, one must think. (A) The days are shorter; (B) the nights are longer; (C) the weather is colder; and (D) sunlight somehow doesn't feel as intense. How large must a solar collector be to gather light from the sun that is low in the sky to collect energy to get through long cold nights? Is there any way to use the summer sun to provide winter heat and winter electricity? And, if so, why isn't such a system available?

Questions Blowing in the Wind

Wind is yet another a form of solar energy with which humanity has had long experience. The newspapers now carry frequent stories about wind farms that are being built around the country. Did we give a Wind Depletion Allowance to Exxon?

Why not put wind turbines on buildings in The Windy City? Is there some good reason why it hasn't been done?

Farmers are struggling to make ends meet (so what's new?). Can they make a better living selling electricity from wind turbines?

Questions about Hydropower

Especially in New England, our forefathers ran factories and fabric mills on hydropower, another manifestation of solar energy. Might there be some reason that hydropower has diminished in importance? (Hint: Do you have a waterfall in your backyard?)

Hydropower can be used to heat water for coffee. That is, descending water is the source of energy that raises the temperature of the water. But we know that water falling over a waterfall doesn't boil when it hits the ground. Doesn't this suggest that a lot of water has to go through a hydropower station to boil a little water?

Questions about the Oceans

The inquisitive reader is aware that the world uses very little energy obtained from ocean waves and tides, but also knows that engineers and entrepreneurs are fully aware of that energy. Are we avoiding the ocean's energy because Big Oil has bought the seashores, perhaps?

Doubtful.

Questions about Firewood

Firewood obviously had a running start, yet all biomass combined provides less than 3.2% of our energy [see Table A11] in the US, and most of that is used for heating. According to *Split Wood Not Atoms* bumper stickers, *some* people are using firewood, but the usage is not terribly widespread. *You,* dear reader, may not be using firewood. Are we to believe that *you* have sold out to Big Oil?

Paper comes from wood. Is anybody suggesting that there is more energy in scrap paper than there was in the wood from which it came?

And what about the frequent news reports about the disappearance of forests? Does this news imply that forests do not renew themselves fast enough to keep up?

Questions about Geography

We have all seen people put their finger on a map of the desert Southwest of the US and sagely tell us that some small part of that territory "could" supply the US with enough solar energy for everything we do.

Just suppose, for the sake of argument, that this optimistic desert picture were true. How would the energy be transported to where it is used?

> "Installing pipes to distribute hot water (or compressed air) tends to be considerably cheaper than installing equivalent electrical distribution capacity."
>
> Amory Lovins (1977)

Even if this bizarre statement weren't blatantly false, one would wonder how hot water or compressed air would run our lights, television sets, and computers.

We have ways of transporting fuels and electricity, of course, that are in common use. We use supertankers (which don't float very well in the desert), pipelines, railroads, highways with tanker trucks, and high-voltage transmission lines. Typically, a large power station sends its power out on HV transmission lines in several different directions, each delivering about 250 MWe to 300 MWe of electrical power. The US uses the electrical output of about 350 such power stations, on the average, not counting peak demand.

Would society really tolerate a distribution system that brought all of our energy from the desert? (Hint: When was the last time they put a pipeline through *your* neighborhood?)

One would naturally expect the sunniest countries to lie in the tropics. Those countries could presumably be able to export solar energy to the US and Europe and make a handsome profit for themselves. But the astute individual knows that the US does not import solar energy from Ethiopia, Kenya, Sudan, Sumatra, or northern Australia. Moreover, the people in those places do not provide their own energy from sunbeams. How, one must ask, is northern Minnesota supposed to be able to do what can't be done in Peru?

Questions about Science and Engineering

Many people evidently see solar energy *only* as a political or economic question. Some imagine, perhaps, that we simply lack the political will to make solar energy happen. Some others think that if we would just throw some money at the problem, we'd become a solar nation. There is an oft-repeated adage that if we would give Exxon a solar-depletion allowance, we'd be using solar energy tomorrow.

However, solar energy is — first and foremost — a topic of science and engineering. It is worth exploring how solar energy actually works in all of its various manifestations. The subsequent chapters deal with those questions.

Chapter 5 discusses the light that arrives at the earth from the sun. We have only one sun, and all of the solar energy manifestations ultimately depend upon that light. It is worthwhile to understand sunlight itself.

Chapter 6 deals with conservation and efficiency. These topics are related to solar energy only in that they can improve the use of solar energy.

The oldest use of solar energy is the burning of firewood. We have already seen that firewood is responsible for about 3 percent of the energy used in the US. Chapter 7 discusses the production of biomass from sunlight, and the production of ethanol from corn, a topic that often finds its way into the evening news.

Hydropower is the subject of Chapter 8. Every time we pick up a newspaper or the utility bill, we are reminded of wind power, the subject of Chapter 9.

When we put up a dam to produce hydropower, the water that is responsible has been evaporated from all over the earth and has fallen as precipitation somewhere in the collection area upstream from the dam. Similarly, the winds that drive our wind turbines picked up their energy from remote places. However, when we use solar energy directly for heat (Chapter 10) or for the production of electricity (Chapter 11), we must collect the solar energy ourselves with devices that we have manufactured and placed in the sunlight. That is, the collection area is precisely as large as our collectors.

There are some other renewable energy sources as well, including ocean waves and tides, and geothermal energy. Ocean waves obtain their energy from the wind, itself an indirect manifestation of solar energy. The tides ultimately obtain their energy from the rotation of the earth, through gravitational force gradients. Geothermal energy is due to radioactive materials that are dispersed throughout the body of the earth. All of these sources will be discussed briefly in Chapter 12.

Chapter 5. We Have One Sun

At first blush, solar energy is both promising and confusing. Both the promise and the confusion come from the large number of manifestations of solar energy. Sunlight grows plants that can be burned for heat, but sunlight can also produce electricity through the use of photocells. Sunlight can heat our homes directly, but it also evaporates water that eventually gives us hydropower. Sunlight drives the winds whose energy can be harnessed with wind turbines, and the winds drive waves whose energy can also be extracted.

The promise is that there are so many ways to use solar energy. The confusion comes when we attempt to compare the various sources. How, for example, can one compare a wind farm with a tree farm? More confusion arises when we compare source with demand. How much hydropower corresponds to the fuel consumption of a fleet of taxis?

Confusion also enters when we listen to engineers, scientists, contractors, farmers, hydrologists, and photovoltaic experts. They all seem to be speaking different languages from one another. Unfortunately, they are, and there is a serious communication problem.

The purpose of this chapter is to establish a common language for all solar-energy topics. The main elements of the language are simply units of measurement that are understood in all corners of the globe. But measurements *always* imply that we will be using numbers. In this case, we already use the same language as everybody else — the familiar decimal system with which we count marbles.

Numbers Large and Small

Most subjects in science are inherently quantitative, and energy is no exception. Numbers and comparisons are inevitable. Many, if not most, people are very uncomfortable with numbers they can't easily visualize. For example, the US uses (in round numbers) about 100,000,000,000,000,000,000 joules [watt-seconds] of energy every year. The energy involved in the formation of one molecule of CO_2 from carbon and oxygen is in the range of 0.000 000 000 000 000 000 1 joule.

Scientists and engineers go out of their way to construct models to aid in visualizing such numbers. For example, the prefix *exa* abbreviated E is 1,000,000,000,000,000,000 (1, followed by 18 zeroes). We represent this large number in *scientific notation* by 10^{18}, which means 10 multiplied by itself 18 times. It is then fairly easy to remember that the US uses about 100 exajoules (100 EJ) of energy every year.

The joule is the international unit of energy. Although it rarely appears in the news media, the unit is understood by every practicing scientist and engineer in the world who was educated since the 1960s. Lift your set of keys about a meter (or a yard), and you have done about 1 joule's worth of work. Becoming familiar with the joule may seem to be too much effort. Rest assured that *not* learning about it requires much *more* effort.

A *joule* is the same as a watt-second, the energy used by a 1-watt lamp during one second, or by a 100-W lamp during 1/100 second.

Complications from Weird Solar Units

People like to have numbers placed in context, and that is one reason writers will tell us that the axis of a wind turbine is 150 feet above the ground, or that the lake behind a dam is 200 feet deep at the deepest end, and is 25,000 acres in extent. In this book, I will occasionally make similar references.

Calculations, however, require something better. The most fundamental question in arithmetic is whether one number is greater than another. A profusion of units often makes such a comparison virtually impossible, leading to an *apparent* complexity of solar energy.

Take a look at the two problems below.

Solar energy problem 1. Which is the biggest?
- A.   11,700 calories per square centimeter during one month
- B.   254 BTU per square foot per minute
- C.   2 MWe generated per 130 acres of solar collector
- D.   1/2 cord of white oak per acre per year
- E.   397 Langleys per day

Solar energy problem 2. Which is the biggest?
- A.   100 W/m^2 or
- B.   673.2 W/m^2
- C.   23.7 W/m^2
- D.   200 W/m^2
- E.   0.001 W/m^2

Problem 2 is a trivial matter for anybody old enough to understand decimal points, but problem 1 requires considerable effort to find tables of the correct data, followed by a lot of arithmetic. It may not even be obvious that the choices in Problem 1 even involve the same type of quantity. That is, are we comparing apples and oranges? In fact, all ten choices for the two problems, in fact, have the same dimensions as solar intensity: energy per unit time per unit area or, equivalently, power per unit area.

Many archaic units like those of Problem 1 exist in the solar-energy literature. It becomes an unpleasant chore to compare numbers found in one source with those in another. A reader must rely on table after table of conversion factors to make sense of the written page. On the other hand, the comparisons become trivial when a consistent set of units is used.

Let's pursue this matter. The following energy quantities are in common use. Here I show the International System of units in italics:

1. Energy

Joule, erg, BTU, calorie, kilocalorie, kilowatt-hour, gigawatt-year, horsepower-hour. Quasi-units: barrel of oil, ton of TNT, ton of coal, gallon of gasoline, cord of wood, ...

2. Time

Second, minute, hour, day, week, month, year, century, ...

3. Area

Square meter, square centimeter, square inch, square foot, acre, hectare, square mile, ...

Solar intensity can be expressed by any one of the energy choices (1) divided by any of the time choices (2) divided by any of the area choices (3), resulting in 728 different units. To make a single table to convert any one of these bizarre units directly to any other requires about a half-million conversion factors. And we're underestimating the situation, because people regularly use about 50 different energy units instead of the 13 shown.

Not every such unit, of course, is in common use. However, firewood production is reckoned in cords per acre per year. People will write that the US annual "energy" usage (yet another unit, but not a constant one) could be generated on 6000 square miles if solar collectors were 100% efficient (unit = US annual energy per year per 6000 square miles). Or, perhaps, a writer may say that if we devoted a mere 20,000 acres of desert to collecting solar energy, we could avoid using 50,000 gallons of gasoline per month in the summer (unit = gallons of gasoline per acre per month). Perhaps the writer is enthusiastic about solar energy, and perhaps he thinks solar energy is a joke. How can the reader gain any perspective when these numbers are so hard to compare?

The United States continues to use a variety of arcane units with equally arcane number systems. For example, we use base-16 arithmetic (16 ounces per pound, for example), and base-12 arithmetic (dozens of this, gross of that). But we also use mixed-base arithmetic (12 inches per foot, but inches divided into sixteenths) and base-60 hybrid systems (60 minutes per degree, 360 degrees per full circle). We use miles of 5280 feet length, and acres, defined so that there are exactly 640 acres per square mile.

A base-10 numerical system has the advantage that children learn to count on their fingers, but children could equally well learn to count the spaces between the fingers[21], as is the habit of cultures that have developed the base-8 system. Computers use the base-16 system. The real advantage of the base-10 system is that nowadays everybody in the world uses it.

Similarly, we *could* redefine our time units so that there were (say) 10 billion seconds in a year or 100,000 seconds in a day, but through long usage, we are content with 60 seconds in a minute, 3600 seconds in an hour, and 86,400 seconds in a day. If every country had its own definitions, we would probably expend some effort to adopt a convenient universal time system. Instead, we use an inconvenient, *but universal* time system.

Historically, there have been thousands of units for measuring length volume, and weight, the units used in commerce. To simplify matters, as well as to keep people honest, the world has adopted universal units for length, volume, and weight, but also for everything else that we measure. If we use those units we are able to communicate with everybody everywhere.

The International System of Units (SI)

During the past century, the "metric" system has evolved from two different but related systems into one, the International System of Units (SI, standing for the French *Systéme Internationale*). The reason most often cited for going metric is that the English system is crazy with its threes, twelves, sixteens, and other numbers like 5280 and 1760. The SI system has tens.

The British have long since abandoned the British measurement system, as have the Canadians, the Australians and all other former British colonies. All scientists and engineers who received their degrees in the US since the 1960s have been trained in the SI system, but have often been forced by circumstance to become mired in BTUs, feet, furlongs, horsepower-hours, acres, leagues, slugs, poundals, and ounces.

---

[21] You can place sticks between the fingers, making it easy to count relatively large numbers.

Yes, the British system is cumbersome, but that's just the beginning. A better reason is that there are no units in the British system for electricity whatsoever.

The most important reason, however, is that we have to make comparisons. Can you tell at a glance whether 7 therms contains more or less energy than 364 gallons of gasoline? How many foot-pounds are there in a BTU? But any fool can tell that 15 of something is more than 10 of something when they are in the same units. That is the best reason for using the same units as everybody else in the world: *The International System of Units*.

You wouldn't know it from the way we behave, but the United States has adopted the metric system — many times. Here is a brief chronology of the metric system.

1585   Simon Stevin suggested that a decimal system should be used for weights and measures, and coinage. Moreover, he suggested measuring angles in a decimal system (as is now done in a number of countries where a right angle has 100 grads instead of 90 degrees).

1790   Thomas Jefferson recommended both a decimal currency and decimal measurement.

1792   The US Treasury produced the world's first decimal currency. (It only took until 2001 for the decimal system to hit the stock market.)

1821   John Quincy Adams wrote a report suggesting the adoption of the metric system. He concluded the report, however, with the recommendation "this may not be the right time to change." This excuse has been used repeatedly since then.

1866   The metric system was made legal (but not mandatory) in the United States by the Metric Act of 1866 (Public Law 39–183). (The English system has never been legalized.)

1875   The U.S. joined 17 nations to form the Treaty of Metre.

1889   The U.S. received a prototype meter and kilogram to be used as measurement standards. That is, we have a metal bar with two scratches that are exactly one meter apart. We also have a standard 1-kg weight[22]. [Nowadays, we define the meter in terms of the wavelength of light from certain atoms. The weight standard remains a certain 1-kg piece of metal.]

1893   The US officially adopted the prototype standards in lieu of its previous length and weight standards. It is of interest that there

---

[22] In 1899, the kilogram was a unit of weight, not mass.

is no such thing as a standard bar for the inch, foot, or yard. Our English units are defined in terms of the metric standards. The inch is defined to be 2.54 cm (0.0254 m) *exactly*. Similarly, our standard pound is defined to be 1/2.205 of a kilogram *exactly*.

1975   The Metric Act was signed into law making metric the preferred system in the US, but it set no target dates.

1988   President Reagan signed a law making all new Federal projects metric, and required all federal agencies to be metric by the end of fiscal year 1992. (The Department of Energy routinely ignores the law.)

1991   President George H. W. Bush signed Executive Order 12770, directing all executive departments and federal agencies implement the use of the metric system.

Watts and Joules

The Energy Information Agency (part of the Department of Energy) uses a bizarre array of energy units in its *Annual Energy Review*, but still a small fraction of the fifty or so energy units in common use. So do the folks who assemble the *Statistical Abstract of the United States*. You need a calculator and a table of conversion factors just to read their tables and compare what's on one page with what's on another.

Never mind that the US government pays no attention to the US government, the international unit of energy is the *joule*.

The energy required to heat a cup of water by one degree Fahrenheit is about 1000 joules. A shotglass of gasoline will produce about two million joules of heat when burned.

Let us emphasize. The joule is the unit used for the quantity of energy. The watt is the unit for the rate of energy conversion. The relationship between the two is expressed as

A joule is a watt-second.

and

A watt is one joule per second

The Underlying Simplicity

There is an underlying simplicity to solar energy. Regardless of whether we are talking about home heating, photovoltaics, biomass, arrays of mirrors, or thermal-gradient ponds, there is something universal to all: sunlight. We have but one sun, the same one our forefathers had. The sun is no different now than it has been for millennia, except for minor changes in output.

It is true that we now know more ways to use solar energy, and that we have even improved the efficiency of various traditional solar systems. But that doesn't make things complicated. A nineteenth-century physicist would not be amazed that we can convert sunlight to electricity, though he might wonder what we intended to do with the electricity. He already knew how to generate electricity with thermopiles. R. E. Day discovered the photovoltaic effect in selenium in 1878, four years before Edison's Pearl Street Station started producing power for New York City.

But the transcendent issue is sunlight itself, not the technology about how it is used. Suppose that somebody invents a solar-powered device to do something we haven't even imagined yet. The device would be inherently limited by the energy it absorbed from the sun. Solar energy devices do not create energy; they merely transform —*some of* — the sunlight into other forms.

With few exceptions, solar-energy manifestations can all be expressed in the same units as those for solar intensity. For example, firewood production, often expressed in cords of wood per acre per year can be expressed in watts per square meter. Food production is often expressed either as tons of product per acre per year or as food calories per acre per year; obviously they can be expressed in watts per square meter. The advantage of doing so is that it allows us to compare — directly, without have to look things up in tables and do tedious calculations — the actual product with the solar cause.

A hydropower dam uses the water in the dam behind it that has been collected from a huge collection area, and can produce some average power throughout the year. Hydropower, then, can also be expressed with the same concept, power per unit area.

A larger wind turbine can produce more power than a smaller one; however, larger wind turbines have to be spaced more widely. Here, too, the same comparisons can be made. A given land area can produce some average year-round power. It is a simple matter to express the wind turbine farm's output in units of power per unit area, the very same unit as for solar intensity, often called *insolation*. (This term should not be confused with *insulation*.)

Of course, man-made devices all have some given surface area that is made to face the sun (and occupy some area on the earth), and can produce some average power, be it thermal or electrical. Regardless, they can all be expressed in solar intensity units, power per unit area.

When all manifestations of solar energy are expressed in one standard set of units, everything becomes simple. Appendix A in this book contains factors for converting units *to* SI units.

Bad Units Are One Cause of Disagreements

How can it be that some people are extremely enthusiastic about solar energy and others see little hope for it? After all, we have but one sun, and the energy it sends us has been well known and well understood for over a century. The world's present energy consumption is well known; no information is being kept secret from the public.

If the sun is to supply our energy, then the year-round average power consumption should equal the year-round average solar power production. One only needs to look at the numbers.

Unfortunately, the simplicity is usually obscured by a bizarre collection of parochial units. How many days must the sun shine on 274 acres of solar collector to produce the equivalent of 22 therms, of 886,000 BTU, or of 17 barrels of oil?

> "Annual wind generated electricity production in California displaces the energy equivalent of 5 million barrels of oil …"
>
> R. Gerald Nix (1995)[23]
> National Renewable Energy Laboratory

> In fact, SCE's refusal to invest in the grid in Tehachapi is suppressing the growth of the industry in the area that could produce enough energy to electrify over a million homes or enough for 2.5 million people.
>
> Greenpeace (2001)[24]

One particularly misleading unit for power is "the home," and Greenpeace is not alone in using this thoroughly obfuscating unit. They never mention commercial establishments or industries. Nor do they ever say that the power figure they give is the full-tilt power *capacity*, not the expected average power. (As we have seen above ["Efficiency in Wind Turbines" in Chapter 2], the average power for wind turbines in 1998 was

---

[23] "Wind Energy as a Significant Source of Electricity,"
http://www.nrel.gov/wind/atlpap2.html

[24] http://www.greenpeaceusa.org/media/factsheets/windtext.htm

only 23.5% of the capacity.) They systematically fail to mention that the wind turbines provide only *electrical* power, which is nowhere near the total power consumption for the "homes."

Simplicity Through Rational Units

The international system of units (SI, *Systeme Internationale*) uses base-10 arithmetic throughout, except for the historical units of time. In the SI system, the basic unit of length is the *meter* (not the centimeter, and not the millimeter, and not the kilometer), and the basic unit of time is the *second*.

Americans recognize the terms *watt* and *kilowatt-hour* because of their use in electricity. There is, however, nothing uniquely electrical about either one of the terms. It is perfectly acceptable to say that a moving car consumes fuel at the rate of 100,000 watts, even though electricity is not driving the car. The term *watt* is simply the SI unit for power — energy per unit time — regardless of the source of energy.

It is not always clear from context whether the term *watt* is being applied to electricity or to some other kind of power. Power plant engineers use *MWe* to refer to megawatts-electric, and *MWt* to refer to megawatts-thermal. A watt of sunlight has the same heating power as a watt of electricity — one joule of energy per second — but the solar watts are not electrical.

A 100-W lamp draws 100 watts of electrical power. If it runs for an hour, the amount of energy it has consumed is 100 watts times one hour, or 100 watt-hours. If it runs for 10 hours, then it uses 1000 watt-hours, or one *kilo*watt-hour. Your electricity bill is for the electrical energy (expressed in kilowatt-hours) used during the month.

However, neither the watt-hour (Wh) nor the kilowatt-hour (kWh) is the SI unit of energy, because the hour is not the standard unit of time. For better or for worse, the standard unit of time is the *second*, so the *standard unit of energy is the watt-second*, otherwise known as the *joule*. (Therefore, the kilowatt-hour is 1000 watts × 3600 seconds = 3,600,000 joules.)

Let us now revisit the problems with which we began this discussion. Solar intensity is in units of *power per unit area*, for which the SI units are *watts per square meter* (W/m^2) and no other. When we use the SI unit, we are automatically in tune with all scientists and engineers everywhere. *By using one agreed-upon system we bypass all problems of the type in Solar Problem 1 on page 67.*

For sunlight, the "insolation" (solar intensity) values are given in Table 3. The values represent the solar flux in watts that strikes a horizontal surface of one square meter.

Table 3: Solar Intensity Values

Solar Flux	Watts/m^2
At earth's orbit (above atmosphere)	1367[25]
At surface, noon, tropics, clear skies	950
Maximum conceivable 24-hour average, at equator, no clouds, at equinoxes	300
Albuquerque, New Mexico, yearly average	240
US, around 48 states, around-the-year, around-the-clock average	200
Hartford, Connecticut yearly average	160

The solar intensity in the US, averaged over all places and over every second of the year, night and day, summer and winter, is about 200 W/m^2. (See the reference in footnote 25 for city-by-city data.) The yearly average at Hartford, CT, is 20% lower than that value, and the value in Albuquerque, NM is 20% higher. In fact, the average solar intensity is within that ± 20% range throughout over two-thirds of the US.

US Energy from Solar?

The land area of the United States is 3.096 million square miles —about 8 trillion square meters — not including Alaska and Hawaii. Suppose that all of that land could produce energy at the rate of x watts per square meter. How large would x have to be to produce the 101 EJ we use per year?

The answer is easy to figure out. Divide 101×10^{18} joules by the number of seconds in a year — 3.16×10^7 to get 3.2×10^{12} watts. Divide that by the number of square meters, and the result is about 0.4 W/m^2. That is, if every square meter in the US could provide an average of 0.4 watts of thermal power, we could supply all of the energy used in the US continuously.

That number is small compared to the 200 W/m^2 of average solar intensity. Bear in mind, however, that sunlight is not the same as electricity or motive power.

---

[25] http://rredc.nrel.gov/solar/#archived. The value accepted in the 1970s was 1353 W/m^2, about 1% lower than the currently accepted value.

Translating

Berman and O'Connor offer the following description of the LUZ International plant in the Mojave Desert:

> "The largest solar electric generating plant in the world is the 355-megawatt LUZ International 'solar-thermal' plant, located between Los Angeles and Las Vegas, which delivers its power to Southern California Edison. Not a photovoltaic plant, LUZ is a 100-acre field of parabolic [mirror] trough collectors in the Mojave Desert…"

Berman & O'Connor (1996)

The LUZ plant produces 355 MWe of electricity in full sunlight, as we will discuss more thoroughly on page 153. Their implication is clear: 355 MWe of electricity is delivered to Southern California Edison. However, the intensity — 355 MWe per 100 acres — amounts to 877 watts per square meter, just under the intensity of noontime sunlight falling on a surface that faces the sun. For the plant to deliver all that electricity, it would have to be nearly 100% efficient, which it certainly is not. Had the authors used SI units, they should have been able to spot the error in an instant.

The Seasons

The sunlight striking the ground (a horizontal surface) is *much* less intense in winter than in summer. For example, at 40° north latitude (New York) less than half as much sunlight falls on a tennis court between 11:30 a.m. and 12:30 p.m. during the winter as it does during the same hour in the summer, even without consideration of atmospheric effects.

However for a surface that faces the sun, there is very little difference between the amount of sunlight striking the surface for the one-hour period around noon in the summer and the winter. In fact, the difference comes entirely from looking at the sun through more of the atmosphere. If you could look straight up to the sun, you would be looking through one thickness of atmosphere, but when you look at the winter sun, you look at it through the *slant height of the atmosphere.* There is some reduction of intensity, but not all that much at temperate or tropical latitudes.[26]

---

[26]Aden E. Meinel and Marjorie P. Meinel, *Applied Solar Energy* , (Addison-Wesley Publishing Co., Reading, MA, 1976), give the following equation for determining the solar intensity for clear skies, given the angle *z* of the

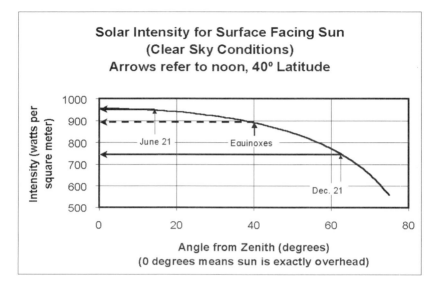

Figure 20: Solar intensity (W/m^2) versus the angle of the noon sun measured from the zenith (directly overhead) for 40° latitude (Philadelphia, Salt Lake City), calcualted from equation in footnote 26. Notice that the vertical axis starts at 500 W/m^2, not zero. The intensity is 890 W/m^2 when the sun is 40° off the zenith, regardless of whether it is late in the day in summer or noon in mid-March.

Figure 20 shows the solar intensity versus the angle of the sun as measured down from the zenith. The arrows in the figure refer to 40-degree latitude, approximately the latitude of New York, Indianapolis, Denver, and northern California. At noon on the summer solstice, the sun is 16.5° off the vertical. At noon on the equinoxes, March 21 and September 21, the noon sun is 40° below the vertical, and at the winter solstice, the sun is 63.5° below the vertical. During the year, the intensity of sunlight at noon varies from a high of just under 950 W/m^2 to a low of about 750 W/m^2. That is, sunlight on a clear day in mid-December is only 21% less than it is in mid-June.

For the two solstices, however, the lengths of daytime are considerably different. Between sunrise and sunset in mid-June, there are 15 hours. In

---

sun, as measured from the zenith: $I = I_0 e^{-c(\sec z)^s}$, where $c = 0.357$, $s = 0.678$, and I_0 is the solar intensity above the atmosphere.

mid-December, the sunlight lasts a little less than 9½ hours. That is, there is 38% less time when the sun is shining.

There is a difference between amount of the sunlight received on a horizontal surface and that received by a surface that is adjusted to be facing the sun at all times. For example, recent several-year average data from Albuquerque (see footnote 25) show that the average solar intensity on a horizontal surface is 233 W/m^2, whereas the average solar intensity on a surface that follows the sun is 366 W/m^2.

If one has one sophisticated two-axis tracking device, the higher intensity is more important. However, if one has an array of collectors in a field, the lower figure is the correct one to use for estimation. The reasons will be made explicit in Chapter 11.

Physical Laws

The laws of physics are not recipes. For example, all airplanes obey the laws of physics at all times, yet the very same laws do not tell anybody the best shape for a wing or the best design for an engine.

But physical laws do provide constraints. That is, they tell us only what we *can't* do.

Perhaps nothing can better illustrate these points than solar energy. The amount of light that comes to us from the sun is (within small variations) constant, the same year after year, century after century. In fact, enough light reaches the *upper* atmosphere every year to melt a layer of ice about 30 meters thick spread out over the entire earth. But when we write down a figure for that amount of sunlight, our ciphers do not tell us how to use all (or any part) of that energy. They do not tell us how to build solar water heaters, photovoltaic collectors, or solar-powered calculators. The fact that we understand how much solar energy impinges on the earth does not tell us how to design anything.

But that figure provides a constraint: the earth cannot receive any more solar energy than the sun sends us. The constraint is just as true for a given parcel of land in a given time interval. The laws of science tell us that no solar collector — no matter how much research is done into improving efficiency — can collect more solar energy than is sent to the collector by the sun.

Chapter 6. Conservation and Efficiency

All of the subsequent chapters in this book are about renewable energy in its various manifestations. Conservation and efficiency, the topics of this chapter, are *not* manifestations of renewable energy. Rather, they deal with how well we use energy.

It is somehow obvious that conservation and high efficiency are desirable goals; indeed there are people who dedicate their careers to improving both. But sometimes what is "obvious" is not necessarily true.

Conservation

The term *conservation* has both a technical meaning and an everyday meaning. Unfortunately, the two meanings are somewhat contradictory.

To a physicist or engineer, the term *conserved* refers to physical quantities that can be neither created nor destroyed. Energy is one of these quantities, along with momentum, angular momentum, electric charge, and a few others.

In common parlance, *conservation* actually refers to things that are *not* conserved in the technical sense. If you drive your car less, you use less fuel. In the everyday sense, you have "conserved fuel." In truth, fuel is destroyed by burning; technically, the quantity of fuel is not conserved. Begging the forbearance of scientific colleagues, I will use the common parlance. I will also refer to fuel and energy *savings*.

That much said, it is understood by all that our fossil fuels are limited, and that we should not be wasteful.

Why bother saving energy?

> Over the next decade or two, Americans should try to cut their per capita energy consumption *in half.* [italics in original.] ...
>
> Paul & Anne Ehrlich (1974)

Saving energy saves money, but is that all?

If the supply of a commodity is infinite, there is little point in saving it. As of press time, there is no known governmental program in Egypt to save sand, or in the Amazon to save water.

If you were given some annual allocation of, say, 500 liters of heating oil (and no firewood or other fuel) every year to get you through the winters in a northern climate, then you should insulate your house very well. You should also be very careful to keep the house just barely warm enough for mere survival, so that you don't run out of oil before the winter is over.

Go back to the same picture, but with the following modification: you are given 500 liters of oil, and *that is all the energy you will ever get.* The 500 liters must last you and all of your descendants *forever.* You may figure out some scheme — using 100 liters per year, for example — that will maximize *your* life expectancy, but it is only a time-buying operation, and leaves nothing for your progeny.

Now look at the worldwide energy picture. In a given year, the US uses one-quarter of the energy consumed in the year, and the rest of the world uses three-quarters. Think of that as four logs on the fire, one labeled US.

Driven by the moral imperatives of the Ehrlics, Amory Lovins, and the Environmental Protection Agency, we want to do the right thing, which is to save energy. However, we will "take it to the max.". *We will cut our energy usage by 100%, and use no energy whatsoever.* We'll take that US log off the fire, and donate it to posterity. We'll store it in New York's Museum of Natural History so that visitors could see our energy supply along with the Hope Diamond, both of which are "saved" and neither of which is "used."

This plan has numerous benefits, aside from making "conservation" zealots proud. There will be no need to pay Social Security benefits to baby-boomers, because they won't reach retirement age. Nobody will ever be around to have to pay off the national debt, so we may let it rise without limit.

According to the commonly accepted scenario, the three logs remaining on the fire will soon be gone, perhaps in three decades, as is commonly assumed. Then the remaining log (retrieved from the museum) will be gone in less than one additional decade. The net effect of our Green zeal — aside from a few hundred million US deaths caused by lack of energy — is a few years of survival for the rest of the world.

And what great benefit would Hillary Clinton's Global Village get if we would cut our energy consumption by only 10% instead of 100%? Less than a year of continued civilization, given the assumption that we'll otherwise run out of energy in three decades.

So, aside from the very legitimate purpose of saving money that might be used for other purposes, saving energy simply buys time — and not much of it — if we indeed would run out of fuels in the immediate future. That is, contrary to popular opinion, conservation of fossil fuels provides no guarantee that they will last very far into the future.

Conservation does make sense when there is some continuing but *feeble* supply. That is the case with solar energy. The total amount of solar energy from all its forms combined — wind turbines, firewood, hydropower, direct sunbeams, and anything else you care to name — is so small that only extreme conservation measures can make solar energy seem realistic.

"THE SOLAR PATH IS SIMPLE: use less energy."

Berman & O'Connor (1996)

With abundant energy supplies — nuclear fission, and (if we can make it work) nuclear fusion — conservation may have its merits, but saving energy resources is not one of them.

Visions of the Future

When I became interested in the societal energy problem in the 1960s, it was clear that writers were concerned only with the remaining years of the 20th century. That limited vision continued until the 1990s. It always struck me as a bit odd, and not only because of the numerological nonsense. (Is there something magic about the number 2000?) We could look back on thousands of years of civilization, but prognosticators acted as if the end of civilization was imminent at (when else?) the End Of The Century. Now that the date has come and gone, people are already beyond that meaningless barrier, if only slightly.

Do we have an energy problem? Our answer to that question depends upon our view of the future.

For example, we might take the apocalyptic view that a giant meteorite will strike the Earth next Tuesday. In this scenario, we have plenty of energy to last to the end of civilization, though that thought is lost on doomsayers.

If we imagine that the hypothetical earth-destroying meteorite will strike the earth later, then we have to analyze just how long various fuels will last. Most commonly, scientists imagine that there is enough petroleum and natural gas to last many decades, but probably not centuries. Coal will last many centuries, but not millennia.

If civilization lasts until, say, 2083, then it is very likely that petroleum will supply only a small fraction of the energy for that year; however, there will still be ample coal reserves. By the fourth millennium, petroleum will be mentioned only in ancient history textbooks, and coal reserves will probably be so depleted as to be considered non-existent, presuming, of course, that civilization lasts that long, and that coal had been used as a major energy source.

Properly used, nuclear fuels are sufficiently abundant to last more or less forever.

To repeat the main point, saving fuel makes the supply last longer. However, *no effort, no matter how grandiose, can possibly save fossil fuels for our descendants 2000 years from now.* (But see A Dissident View, below.) Saving fuel saves money and buys time. No more, no less.

A Dissident View

Thomas Gold, Professor Emeritus of Cornell University, is. a current champion of a different view of the origin of natural gas and petroleum.[27] He argues forcefully, as have others before him, that petroleum did not come from decaying vegetable matter in bygone eons. According to Gold, meteorites have brought simple hydrocarbons to the earth for eons, and bacteriological processes have produced more complex hydrocarbons (*a.k.a.* petroleum), tens of kilometers deep within the earth's crust. The energy source is the same one that has recently been discovered to maintain life in the deep oceans — geothermal energy.

Therefore, nobody has any idea how much hydrocarbon content we have in the earth's crust. *If* they exist only as a fossil fuel, that is one thing; *if* they are the result of cosmic processes, that is quite another. If Professor Gold is correct, then the supply of hydrocarbons is virtually infinite.

Perhaps Professor Gold is correct; perhaps he is making an optimistic error. Funding agencies do not look favorably on such dissident views, so even the necessary research is lacking. Until such enormous sources of energy are actually found, however, we must regard "fossil fuels" as a finite resource whose duration may be decades or centuries, but probably not millennia.

Home Insulation

Our pioneering ancestors understood perfectly well that well-insulated houses were easier to heat than poorly insulated ones. Unfortunately, they had few options. The best ones were tightly caulked log cabins, but they required a lot of wood.

The balloon house was a brilliant invention that sprang up in the mid-1800s in the American Midwest. Before that time, houses were commonly log cabins of the Abe Lincoln variety. Whereas the entire volume of a wall in a log home is wood, the walls of a balloon house were mostly hollow,

---

[27] See http://people.cornell.edu/pages/tg21/usgs.html for Gold's excellent paper on the non-biological origins of petroleum and natural gas.

thereby using much less wood. Balloon houses make efficient use of wood, but not necessarily of heat.

Figure 21 shows the construction of a stud wall of a balloon house. Even with the usual plywood exterior, much less wood is used in the balloon house than in a log house. The balloon house is, therefore, conservative with respect to the use of wood. The price to be paid is that the spaces between the studs need to be filled with some kind of insulation.

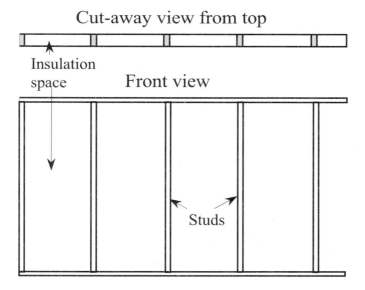

Figure 21: Top (cut-away) and front views of a stud wall of a balloon house.

Insulation

Insulation was a serious problem for balloon houses. Straw was often used because it is a good insulator; its high flammability was one of the hazards of that good insulation. Fiberglass, the standard insulation for balloon houses nowadays, was invented a full century after the balloon house was invented. Styrofoam and its relatives came even later.

Heat loss also depends upon the thermal resistance — the so-called R-value. See Appendix B for details. The important part is that the heat loss through a wall increases with increasing temperature difference between inside and outside, and with the area of the wall. The heat loss decreases as the wall thickness increases.

Given those simple ideas, the National Weather Service keeps track of *degree-days*, as described in the next section.

Degree-Days

One measure of the heating requirements of a given geographical area is the number of *degree-days*. When the average temperature for a day is below 65 °F, it will be necessary to provide some heat for the home. (If the average temperature is above 65 °F, the heat from human inhabitants, the lights, and the appliances is adequate to keep the house warm enough.)

The degree-day is not in the international system of units, for two reasons. First, the temperature is in degrees Fahrenheit (not Celsius), and second, the time unit is the day, (not the second).

The National Weather Bureau keeps a running sum of the difference between 65 °F and the average temperature for each day. For example, if the average temperature for the 7th of November of a given year is 40 °F, then 25 degrees-times-one day is added to the previous total of degree-days. By the end of the heating season, there may be 1000 degree-days in one location and 8000 degree-days in another. The amount of heat that a given house requires annually is proportional to that total.

A similar total is kept for *cooling* degree-days, a sum that determines how much air conditioning a house will need.

For the US as a whole, the Weather Bureau tabulates the US-wide average annual value for both heating degree-days and cooling degree-days. Figure 22 shows the average heating degree-day total for the entire US from 1949 to the present, and Fig. 23 shows the cooling degree-days. Local weather stations keep degree-day information for use by heating and air conditioning contractors. Those values for heating vary from about 1000 degree-days to about 10,000 degree-days (per year), depending upon location.

Global Warming and Home Heating

If the US were experiencing any systematic warming, it would be gradually becoming easier to heat our homes and gradually becoming harder to air condition our homes. In other words, there would be a downward trend for the heating degree-day average in Fig. 22. For example, if the US were to have warmed by 1 °F during that 50-year time span, the number of degree-days would have decreased by about 200, since there are about 200 days in the heating season, averaged over the US. The data show no such trend.

US Heating Degree-Days

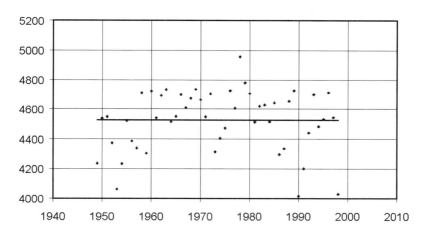

Figure 22: US average heating degree-days, 1949 to present, data from *Annual Energy Outlook*. Notice that the scale does not start at zero, but runs from 4000 to 5200 (°F-days). Warm years are indicated by low heating requirements, *i.e.,* by fewer degree-days. The absence of a downward trend indicates that global warming (if it exists worldwide) does not exist in the US. (Data from EIA's *Annual Energy Outlook 1998.*)

The data for cooling degree-days in the US are shown in Fig. 23 for the period 1949 to the present. Warmer years require more cooling, and the number of degree-days is higher for those cases. Again notice the absence of a trend that would indicate warming. If the US had warmed by 1 °F in the time period, there would be something like 200 more degree-days now than in 1949.[28]

---

[28] The largest contributions to the cooling degree-day total come from the south where the cooling season is longer, just as the largest contributions to the heating degree-day total come from the north where the heating season is longer. For simplicity, I have used 200 days for each.

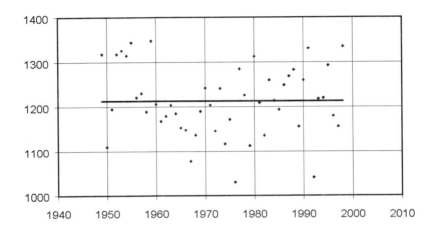

Figure 23: The average US cooling degree-day total for each year since 1949. Note the absence of a trend that would indicate any systematic warming of the US.

Insulation in Solar Homes

Heating bills in almost any home can be reduced by improving the insulation. Doubling the thermal resistance cuts the heat loss in half. Doubling it again can cut heat loss in half again. Some people go overboard in insulating homes, turning them into houses with walls as thick as those of a medieval castle, but insulated with fiberglass and Styrofoam instead of stone. A heavily insulated home requires much less heat in winter and much less cooling in summer.

There are often accounts in the Sunday supplements of new solar homes, complete with photographs and wondrous tales of lower heating bills. Usually, both reporter and homeowner alike attribute the improvement to sunlight. More correctly, most of the *improvement* comes from improved insulation. Most of the *expense* comes from installing solar collectors.

The advantages of good insulation are not limited to saving money. A well-insulated home is not drafty, nor does it have uncomfortable cold spots. For a solar home, superb insulation is an absolute necessity. Otherwise, solar energy doesn't have a chance to heat the home in any cold climate. The insulation is just as necessary in hot climates to reduce air conditioning bills.

Contrary to Popular Belief ...

There are two popular notions about conservation and efficiency that are manifestly incorrect. Let us address them.

What Conservation is NOT

Let us begin with a few assertions commonly made in the news media..

> ➢ Insulation is a good idea.
> ➢ Driving efficient vehicles is a good idea.
> ➢ Using efficient motors is a good idea.
> ➢ Turning off unnecessary lights is a good idea.
> ➢ **In short, saving energy is a good idea.**

That's all mother-and-apple-pie stuff that nobody disputes. But it has become chic to add a totally false proclamation, namely, that conservation is a *source* of energy. I would challenge anybody to demonstrate that assertion by locking himself or herself in a well-insulated empty room, and use every method of conservation, thereby to fill one shot-glass with gasoline.

> "By ignoring efficiency, the Administration may be missing quickest, easiest energy *source*." [emphasis added]
> *Chemical & Engineering News.*[29]

Such nonsense is unfortunately rather widespread. By being efficient, we can use less energy for project X, and have more available for project Y; however, to regard conservation as a *source* of energy is merely to play with words. Paul and Anne Ehrlich are wise enough to recognize the word-play, and assiduously put the word "*source*" in quotes.

There are many good arguments to be made for conservation, but its being a source of energy is not one of them.

---

[29] Jeff Johnson, "More is Better in Bush Energy Plan, *By ignoring efficiency, the Administration may be missing quickest, easiest energy source,*" *Chemical & Engineering News*, p. 37, (May 14, 2001).

Conservation is no more a source of energy
than
Dieting is a source of nutrition

Hungry? Go diet! Consider California's electrical crash-on-the-shoals. Here are comments from the solar sirens.

"Here is a case in point. As of 1975, Pacific Gas and Electric planned nine new nuclear plants and one coal plant to satisfy the growth in demand anticipated by the early 1990s. But the Environmental Defense Fund demonstrated to the California Public Utilities Commission that the company could avoid the need for the new plants and save money at the same time by peddling devices like efficient light bulbs …"

Oppenheimer & Boyle (1990)

So long as California followed my advice to do the cheapest things first, that state had ample electricity …

Amory Lovins (2001)

California found that their low-energy diet did not produce energy. Throughout the 1990s, power demand *increased,* and reserve power *decreased* until there was none. California lived off the fat that others had stored. As they say, it isn't the fall that hurts; it's the sudden stop at the bottom.

What High Efficiency Will Not Do

It is "obvious" to the casual observer that if our machines were more efficient, we would use less energy; however, nothing could be further from the truth.

Imagine, as we discussed above, that our engines were the inefficient Newcomen engines of the mid-1700s, producing 5 joules of work for every 10,000 joules of input heat from coal. How much energy would all of the cars on the road be using today? Very little. Nobody could afford them. The cost of manufacture would be extremely high, and so would the cost of fuel.

The same phenomenon has been true of computers. When many kilowatts of input power to large mainframe computers of the 1960s produced results more slowly than today's handheld flea-power calculators, the total amount of energy used every year for calculating was utterly negligible in the US energy picture. Now that computing efficiency has become extremely efficient compared to that in the past, the major increase in demand for electric power has come from computer-related activities. Peter Huber[30] comments,

"The U.S. today consumes 100 quadrillion BTU (quads) of thermal energy in a year. In 1950 the figure was 35 quads; in 1910, about 7 quads, not counting horses and other agricultural sources.

"The efficiency of energy consuming devices always rises, with or without new laws from Congress. Total consumption of primary fuels arises alongside. The historical facts are beyond dispute. When jet engines, steam power plants and car engines were much less efficient than they are today, they consumed much less total energy, too."

But the efficiency paradox is nothing new. In the 19th century, the efficiency of steam engines was steadily improving as a result of James Watt's steam engine. For a while, the consumption of coal decreased by as much as a third, but in the subsequent thirty-three-year period the consumption increased tenfold. An English economist, Stanley Jevons, commented on the paradox in 1865[31]:

"It is wholly a confusion of ideas to suppose that the economical use of fuel is equivalent to diminished consumption. The very contrary is the truth … It is the very economy of its use which leads to extensive consumption. It has been so in the past and it will be so in the future."

When we find ways to increase efficiency, we reduce our overall consumption, but only temporarily. We soon find ways to use energy that were previously too expensive. When our cars become more fuel-efficient, we drive more. When the Internet becomes faster, we send more information. When lighting becomes cheaper, we tend to light up the outdoors, much to the dismay of astronomers. When refrigeration becomes cheaper, we buy both a refrigerator and a freezer. When hot water is cheaper, we install hot tubs.

---

[30] Peter Huber, "The Efficiency Paradox," *Forbes*, p. 64, August 20, 2001.
[31] Thanks to J. C. Maxwell of New Galloway, Scotland, for furnishing me with this quotation from Jevons.

Summary

Conservation and efficiency have their merits, but the concepts are vastly over-rated.   Even extreme conservation cannot guarantee the existence of a fuel in the future. The effect of increasing efficiency is not to decrease consumption, but rather to increase consumption.

Chapter 7. Biomass

After the OPEC oil embargoes, there was a spate of Split Wood Not Atoms bumper stickers, at least in New England where trees are plentiful. We understand that the bumper stickers didn't sell well in Western Kansas, where early farmers used stone fence posts because there were no trees for miles around.

There is a certain appeal to the notion of energy independence gained by burning wood from local trees rather than oil from medieval sheikdoms. However, the idea is more of a delusion than a dream, for at least three reasons: (1) tree growth could not possibly hope to keep up with demand for firewood; (2) pollution from wood burning is far greater than from burning petroleum; and (3) environmentalists oppose the concept.

Efficiency of Biomass Production

The term *biomass* refers to any matter that recently came into existence as a result of photosynthesis — trees, leaves, grass, corn, sugar cane, and weeds. The category includes secondary products such as ethanol derived from corn, and dung from domestic or wild animals.

We are usually concerned with the *amount* of such matter (produced, say, on a given plot of land during one growing season) usually reckoned in kilograms of *dry matter*. When the water content is removed, the *heat content* of all biomass is about the same, about 15 MJ/kg, which is about 1/2 that of coal and about 1/3 that of petroleum (see Table A3).

Every child learns in grade school — either by reading or by actually experimenting — that sunlight causes plant growth directly. If a portion of a leaf is covered with a dark sheet of paper, the leaf turns yellow in the covered area. In the wild, when one leaf totally blocks sunlight to another, the hidden leaf would wither and die. Ideally, a plant would have just enough leaf area to absorb all of the sunlight, but not so much that it puts out leaves that don't get enough sunlight to survive. The plant must meet those conditions in full sunlight, windy conditions, cloudy weather, at all times of the day, and at all times of the year when the weather is good enough to support plant life.

As it happens, most plants have about five times as much surface area as the land beneath them.[32] That is, the leaf area of the plants covering a hectare (10,000 m²) of ground will have a total surface area of five hectares. But plants don't actually absorb all the sunlight even when growing conditions are ideal. Leaves are, after all, *green*. This means that they reflect green light, approximately the strongest portion of the solar spectrum. Chlorophyll absorbs light only in the blue and red portions of the spectrum, and not completely even there.[33] With no further information, we could confidently estimate that no more than about 20% of the energy sunlight could become stored as biomass energy.

But we do know more. Generally, it requires the absorption of about ten photons of light in the red/infrared range to produce one carbohydrate molecule. "Of the 73×10^8 kcal [30×10^{12} joules] falling on an acre [4047 square meters] per year [3.16×10^7 seconds] in temperate latitudes, only about 0.1 to 0.5 percent is fixed as organic material," says L. F. Small.[34] Small's figures for sunlight amount to 239 W/m², about equal to the average solar intensity in Albuquerque. The rates of fixing organic material, 0.1% and 0.5%, correspond to 0.239 W/m², and 1.19 W/m², respectively.

The most prolific growth of biomass I have been able to find is that of sugar cane grown in a subtropical environment.[35] Alan Lloyd, a professional engineer who has long lived in Hawaii described to me how the sugar cane is harvested. "Ordinary harvesting equipment won't work. They use a big D-8 [800-horsepower] Caterpillar tractor to smash down the sugar cane. Then they use big steam shovels to hoist it onto trucks that haul it away to the sugar factory."

The production of biomass energy in this case amounts to 3.7 W/m².

---

[32] R. G. Loomis, W. A. Williams, and W. G. Duncan, "Community architecture and the productivity of terrestrial plant communities," in *Harvesting the Sun: Photosynthesis in Plant Life*, San Pietro, Greece, and Army, eds. (Academic Press, New York, 1967).

[33] C. T. deWitt, "Photosynthesis: Its relationship to Overpopulation," in *Harvesting the Sun: Photosynthesis in Plant Life*, San Pietro, Greece, and Army, eds. (Academic Press, New York, 1967).

[34] Lawrence F. Small, Professor of Oceanography at Oregon State University, in Douglas Considine's *Energy Technology Handbook* (McGraw-Hill, 1977).

[35] Alfred L. Johnson, "Biomass Energy," Astronautics and Aeronautics **13**:11, p. 64 (1975).

Split Wood, Not Atoms!

Some New Englanders have a long tradition of cutting trees for firewood, *sustainably*, in the current jargon. In order not to deplete the forest, it is necessary to do selective cutting. Every year, an acre (about 0.4 hectares) of land can be farmed for about a half cord of firewood. (A cord is a pile of four-foot [1.2 meter] logs stacked four feet high in an eight-foot long stack, comprising 128 cubic feet [3.6 cubic meters] of wood and air.)

Let's not lose track of the goal, which is to express all solar notions in SI units for easy comparison. The half-cord represents a certain amount of thermal energy — 16 billion joules; the land area is one acre (4047 square meters); and the time is one year (31.6 million seconds). Dividing the energy by the time, we find that our acre of forest produced a mere 500 watts of power (average energy stored per unit time). Dividing the 500 watts by the 4047 square meters yields about 0.12 W/m^2, representing an efficiency of about 0.06% for converting sunlight into chemical energy stored in the wood.

Imagine, now, that the entire United States is covered with trees that produce firewood as abundantly as New England forests. The sustainable yield of firewood would amount to about 970 GW_t (thermal), a mere 30 percent of our around-the clock consumption rate of 3,200 GW_t (3.2 TW_t). Of course, at least half of the US is utterly incapable of producing trees at the prodigious rate of New England forests. Helping to make our case for us are Paul and Anne Ehrlich, who say,

> "Humanity now obtains about 1.5 $TW[_e]$ [terawatt, 10^{12} watts] by using biomass fuels—fuelwood, crop wastes, and dung." [referring to heat, not electricity] … Burning residues or dung is a desperation measure taken by destitute people lacking other fuel sources. Using these materials as fuel leads to a steady depletion of soil nutrients and fertility, causing a stark deterioration of farmland…
>
> Paul & Anne Ehrlich (1991)

In other words, the world's entire usage of biomass energy is less than half of the energy consumption of the US alone, and a mere 12 percent of the worldwide energy consumption. That is, on the average, the world uses about 13 trillion joules of energy every second, but collects only 1.5 trillion joules from biomass in the same second.

Ethanol From Corn!

One of those manifestations of solar energy is ethyl alcohol (*a.k.a.* ethanol, *a.k.a.* EtOH) from corn, known as "liquid sunshine" to the Solar Pollyannas. Ethanol is the same alcohol (CH_3CH_2OH) that is in beer, wine, and distilled spirits. The process of producing EtOH from corn begins with using the very same yeast that consume sugar and produce EtOH as a byproduct.

As they produce more and more EtOH, the yeast are living in their own offal, so to speak. After the concentration of EtOH becomes higher than about 10% (depending upon the strain of yeast), the yeast can no longer function; in fact, they die.

On the early American frontier, the "proof" of good liquor was that it was flammable. As it happens, a mixture of water and EtOH will not burn until the concentration of EtOH is 50%, well beyond the capabilities of yeast. For this reason 100-proof became associated with 50% concentration. (Pure EtOH is 200-proof.)

The 90%/10% mixture of water and EtOH produced by yeast will not burn; it is not a fuel. Therefore, something needs to be done to the "mash" left after the yeast have had their day. That something is distillation — boiling the mixture and then re-condensing the EtOH-rich vapor — repeatedly until the concentration of EtOH is high enough. (It is impossible to get beyond 190-proof without adding benzene, which must then be removed if the EtOH is to be drunk.) The amount of energy that must be used for distillation is obviously large. How large? Read on.

The Ethanol Lobby

In gasoline, EtOH increases the octane rating. The Environmental Protection Agency in the Clinton years ruled that gasoline manufacturers were required to add EtOH into gasoline as an oxygenate; moreover, the EtOH was *required* to be derived from corn. The oil companies sued the EPA in federal court. The court ruled that the EPA had the right (on grounds of protecting air quality) to establish the EtOH content of gasoline, but did *not* have the right to specify the source. In truth, EtOH is obtained more cheaply from petroleum than from biological sources like corn.

Archer Daniels Midland, the major corporate entity behind EtOH, is obviously part of the ethanol lobby, as their main concern is farming and farm products. But there are also manufacturers of equipment for distillation and transportation of the EtOH product.

How much subsidy does ADM get? According to James Bovard,[36] "Every \$1 of profits earned by ADM's corn sweetener operation costs consumers \$10, and every \$1 of profits earned by its ethanol operation costs taxpayers \$30."

In 1997, I had some communications with a manufacturer[37] of distillation equipment for the production of ethanol from corn. He was unwilling to divulge how much energy actually was used in the production of EtOH from corn. In fact, he was very evasive and defensive and refused to provide any data at all. He was obviously trying to hide and/or make excuses for very low (or even negative) energy production. There was no point in trying to nail that custard pie to the wall.

In the Still of the Night

Fortunately, honest data can be obtained from people who are interested in facts rather than in preserving their financial interests.

According to Cornell Professor David Pimentel[38], there is a net energy *loss* of about 18.9 MJ for every kilogram (given by him as 54,000 BTU for every gallon) of EtOH. *That is, it takes more energy to produce EtOH than you can get out of the EtOH.*

More recently Pimentel[39] chaired a US Department of Energy panel to investigate the energetics of ethanol production. They found that "131,000 BTUs are needed to make 1 gallon of ethanol. One gallon of ethanol has an energy value of only 77,000 BTU. ... there is a net energy loss of 54,000 BTU [per gallon]."

He continues, "That helps explain why fossil fuels — not ethanol — are used to produce ethanol."

But there are other costs as well. Erosion of soils occurs 12 times as fast with corn growing as with other crops, and corn requires 25% more water.

Ignoring the minor matter of negative efficiency, Pimentel goes on to say, "If all the automobiles in the United States were fueled with 100 percent ethanol, a total of 97% of the U.S. land area would be needed to grow the corn feedstock."

---

[36] http://www.cato.org/pubs/pas/pa-241.html James Bovard, "Archer Daniels Midland: *A Case Study In Corporate Welfare* (Policy Analysis #241, Cato Institute, September 26, 1995).
[37] Name omitted to protect the guilty.
[38] David Pimentel, "Energy and Dollar Costs of Ethanol Production with Corn," Hubbert Center Newsletter #98/2 (Apr. 1998).
[39] http://unisci.com/stories/20013/0813012.htm

Pimentel's data are somewhat naïve, according to some experts, primarily because he did not include the energy value inherent in the dried mash. Those who wish to see details should consult the thorough Shapouri paper[40] on the net.

Shapouri *et al.* use the following data: The heat value of EtOH = 76,000 BTU/gallon; corn is produced on good farms at 122 bushels per acre per year; and one bushel of corn yields 2.55 gallons of EtOH. They use data from the literature to account for all possible inputs, losses, and residuals to arrive at the following statement: "...for every BTU dedicated to producing ethanol, there is a 24-percent energy gain."

I do not intend to challenge these figures; in fact, the background research seems to have been very carefully done. I do, however, intend to put their numbers into perspective by converting to SI units for solar intensity — watts per square meter.

Readers not interested in the arithmetic may skip forward to the results in the next paragraph. Multiplying 76,000 BTU per gallon by 2.55 gallons per bushel, and then by 122 bushels per acre per year yields 23.6 million BTU of EtOH energy per acre per year. Multiplying by 1054 J/BTU to convert the energy to joules, and dividing by the 31.6 million seconds in a year yields a year-round average gross power of 789 watts per acre (0.195 watts per square meter). If these results are not low enough to be discouraging, let us remember that these values are only 24% greater than the average power input to produce the EtOH. That is, 600 watts per acre is the annual average power required from fossil fuels to produce 789 watts per acre.

Finally, the *net* around-the-clock average power available from EtOH production from corn amounts to a paltry 189 watts per acre, or 0.047 watts per square meter. If corn-derived EtOH were used to produce electricity at an efficiency of 33%, one acre's worth of corn could keep a 60-watt light bulb burning continuously. We remind the reader that Shapouri *et al.* use the most optimistic figures: the *best* corn yield; the *least* energy used to produce fertilizer; the *least* energy required for farming; the *most efficient* distillation techniques; the *most* residual energy (in the form of mash); and generally *the most favorable* (but still credible) values for any and all aspects of EtOH production.

Suppose that we could establish huge farms for EtOH production to supply the entire US with energy. Ethanol production yields a net 0.047

---

[40] Hosein Shapouri,, James A. Duffield, and Michael S. Graboski, "Estimating the Net Energy Balance of Corn Ethanol" (paper AER-721): www.ethanolrfa.org. (rfa = Renewable Fuels Assoc.)

W/m^2, some 4300 times smaller than the 200 W/m^2 average US insolation (intensity). All we would need would be nearly seven times the land area of US, devoted to EtOH production, using the most efficient methods on the planet, with no land set aside for cities or National Parks, to produce the energy used in the US.

Maybe we can buy Russia, China, Canada, Brazil, …

Manure!

Chicken farms, dairy farms, and feed lots produce prodigious quantities of manure. It is possible to process the manure to produce methane, the principal component of natural gas. Getting rid of manure and producing energy while doing so are both laudable goals.

Typical of articles found in the popular press is one in *The Hartford Courant*[41], in which the electrical power output from the methane works out to about 100 watts per head. Another is from the Environmental Protection Agency (see Ref. 50), according to which

> "Dairy cows provide the major recoverable animal manure resource in Washington. In 1992 the manure generated by about 242,000 dairy cows had the potential to produce 26 aMW of electric power. The generating potential is based on the rule of thumb of slightly more than 0.1 aMW per thousand head of dairy cows."

And what's *aMW*? It's *alternative* megawatts. Note again that 26 million watts divided by 24,000 cows is 107 watts per cow. Similarly, a farmer in Durham, California[42] is producing 40 kW from the manure produced by 320 cows, amounting to 125 W/cow.

In fertile places where one acre of well-fertilized land can support ten head of cattle (exclusive of the grain fed at milking time), these figures amount to about 1000–1250 watts per acre, or about 0.25–0.3 W/m^2. Sunlight stores energy in vegetable matter, cows use some of that energy, the methane digester uses some of the energy that is left in the manure, and the methane contains the residual. Heat lost in the production of electricity consumes even more. The overall efficiency of converting sunlight to electricity is low, as one would expect, around 0.15% at best.

---

[41] *The Hartford Courant* [5/13/97]
[42] http://www.jgpress.com/BCArticles/2001/030146.html

Expectations

Of course, biomass does have its place in energy production. We in the US obtain a little over 3 percent of our energy from biomass.

One way to increase the production of biomass energy would be to convert farmland from the production of wheat to the production of some hypothetical energy crop. That's a pretty unlikely scenario, given that people like to eat.

Another would be to use more fertilizers.

Another way would be to irrigate the deserts of the world with desalinated seawater. But the energy requirements for desalination could not be met by solar systems, least of all by the power generated from the biomass grown from the newly irrigated land. Nor could solar systems provide the power required to pump water up[43] to high elevations like Utah, Nevada, New Mexico, and much of the Great Plains.

We note for the record that environmentalists have expressed their opposition to all of these methods.

That is, there is very little likelihood that biomass energy production can increase appreciably beyond its present production, and environmentalists would obviously oppose devoting the entire US land area to production of energy crops.

---

[43] In hydropower plants, water descends and produces electrical power. To use electrical power to pump water to a higher elevation should be called *inverse* hydropower.

Chapter 8. Hydropower Is Solar Power

Sunlight evaporates water. Clouds form. Precipitation falls. Water runs downhill. Hydropower plants use the weight of water behind dams to turn turbines to generate electricity. Hydropower, therefore, is solar power. It is efficient — over 85% for large power plants — and is as renewable as the rains. It's just the ticket for environmentalists.

Or so you'd think. During the Clinton Administration, EPA Head Carol M. Browner removed hydropower from the list of renewable energy sources. It may be renewable, and it may be solar, but hydropower is no longer politically correct. Hydropower — at least from dams large enough to produce a respectable amount of power — is no longer considered "Green." In this case, at least, environmentalists admit to the hazards aspects of solar energy.

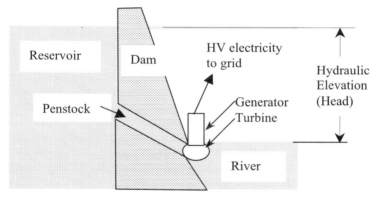

Figure 24: Generation of hydropower. Pressure due to the water behind the dam forces water through the turbine, causing it to rotate, in turn rotating the generator. The amount of power produced depends upon the volume flow rate through the turbine and the elevation of the reservoir above the river below the dam.

Figure 24 is a sketch showing the water flow in a typical hydropower plant. Water flows through the turbine, which rotates the shaft of the generator, producing electricity.

The amount of power available depends upon the volume flow rate of the water (cubic meters per second in SI Units). It also depends upon the elevation of the water surface in the reservoir — variously called the *hydraulic elevation*, the *head*, and the *pressure head* — above the river below the dam.

Hydropower plants can be turned on in a relative hurry, as there is no waiting time to heat up the system. They are therefore the preferred source of electricity for peak power, for example, during the hot afternoon while air conditioners are working very hard.

In July, 2001, the mayor of Seattle, Washington, was upset that President Bush was not willing to sign the Kyoto accords (that the Senate had already rejected 95:0 during the Clinton administration). To declare the city's CO_2 purity, the mayor declared that the city would begin immediately to reduce its CO_2 emissions. "We have plenty of hydropower," he said. (Grand Coulee's operators, whose electricity the good mayor is co-opting, must be wondering "Who's this *we*?")

Let's have a further look at this mayoral buffoonery. It is axiomatic that the operators at Grand Coulee do not waste any of the energy available at the site. To do so would be to throw away money. The only thing Grand Coulee needs to produce more energy is more water in the reservoir, and the good mayor's rain dances haven't been particularly successful. Grand Coulee produces all of the energy it can, and all that energy offsets the consumption of fuel somewhere. However, there is no *spare* energy to offset the consumption of *more* fuel, thereby to *reduce* CO_2 emissions in Seattle or elsewhere, even if the good mayor lays claim to the emission-free electricity.

Hydroelectric power plants produce high power for short times. While a hydro plant is producing power, the water level drops in the lake, because the flow rate through the turbines is usually greater than that of the streams that feed the lake. When the generator is turned off, the water level rises. Hydro plants are designed so that their year-round average power output is commensurate with the year-round water flow.

Large coal-powered and nuclear power plants typically produce 1000 MWe to 1200 MWe. The capacity of the very impressive Hoover Dam (*nee* Boulder Dam) on the Colorado River is 2079 MWe. (A picture of the somewhat smaller Glen Canyon Dam is shown in Fig. 25.) During 2000 (a typical year), Hoover produced 5.29 billion kWh, which means that its *average* output power was 604 MWe. Its *capacity factor* was 29.1%. That is, it produced 29.1% as much energy as a 2079 MWe power plant would produce if run at full power around the clock for a year.

The Hoover Dam collects the precipitation from a land area of 430,000 km², 60% larger than the land area of Colorado, storing it in Lake Mead. The total storage volume is 35 billion cubic meters. The *hydraulic elevation* — the height of the surface of the lake over the river below — is 175 m (576 feet).

Figure 25: The Glen Canyon Dam on the Colorado River. The capacity is 1,296 MWe, and the average power output for a ten-year period is 570 MWe (capacity factor is 44%). Picture from http://www.nrel.gov/data/pix/searchpix.html, Larry Gordon, Bureau of Reclamation, photographer.

Hydropower in the US Energy Picture

It is interesting to look at the average power output per unit area, the so-called *energy intensity*. That is, Hoover Dam produces an average of 604 megawatts, but the water comes from 430,000 square kilometers. How much is that, expressed in watts per square meter of land collection area? An incredibly small 0.0014 watts per square meter. Of course, getting hydropower from the runoff and using the land for other purposes are not necessarily mutually exclusive.

The Hoover Dam's collection area includes the arid Southwest. For the US as a whole, the capacity of all existing hydropower plants combined is 97,500 MWe, and their average electrical power production is 37,500 MWe. The average power intensity — 37,500 MWe divided by 7.68 trillion square meters of US land area — is 0.0049 W/m^2, nearly three times that for Hoover Dam.

The energy intensities (either the 0.0014 W/m^2 or 0.0049 W/m^2) of the previous paragraphs should not be misinterpreted. The figures refer to the approximate amount of power that can be obtained from a huge tract of land that drains into a reservoir for a hydropower station. They specifically do not refer to the amount of land that is occupied by the reservoirs themselves.

The Bureau of Reclamation oversees the operation of 58 hydropower dams in the West, of which Grand Coulee is the largest. Its capacity is 6809 MWe, and its net generation in 2000 was 22.8 billion kWh, amounting to an average power of 2,600 MWe. Its capacity factor was 38%.

It is worth repeating that the low capacity factors for hydropower does not imply that these power plants are undependable. They are designed to produce lots of power for several hours per day when needed, but not continuously. The maximum power available — the capacity — is determined by the generators, but the year-round average power is determined by the annual rainfall.

The first large hydropower plant to be built was Niagara, which produced power first in 1882. It is obvious that engineers would first harness power from sites that offered a great deal of potential, as indeed they have. It becomes less and less cost-effective to build smaller and smaller dams. I regard it is highly unlikely that the electrical output of hydro in the US could ever triple the current output.

Example: A Local Non-Solution

The annual rainfall in Connecticut, a state of about 13,000 km^2 (5000 mi^2) is about 1.1 meters (44 inches). Imagine collecting *all* of that water behind a dam of about 60-meter (200-foot) height. Doing so would inundate well over half the state, but we're just doing a calculation here, not planning to drown the Governor. (It would take years, of course, to fill the lake from rainfall, but we'll ignore that little problem.) How much around-the-clock power could we get?

About 225 MWe. Unfortunately, the state uses about 3,500 MWe around the clock. In other words, flooding over half the state with a dam holding water to 60-meter (200-foot) elevation would produce only about 6.5 percent of the state's demand for electricity.

Worldwide Hydropower

Most of the world lags behind the US in energy production. All of our large hydropower sites are already in use, but elsewhere, there is power to be exploited. The Three Gorges project,[44] will be the world's largest hydropower station when it is completed in 2009.

The Yangtze River is 5989 km long, behind only the Nile and the Amazon in length. In terms of water flow, it ranks only behind the Amazon and the Congo.

The Three Gorges Dam will have twenty-six 700-MWe turbines for a total power output of 18,200 MWe, which is bigger by half than the world's present leader, Itaipu Dam on the Brazil-Paraguay border. The dam will stretch about two kilometers and be 175 meters high. The lake will be 600 km in length and will flood 28,000 hectares (280 square km, about 100 square miles).

In addition, the dam will serve to control floods (two floods in 1931 and 1935 combined to kill nearly 300,000). But Lovins is more interested in the environment than in saving human lives.

> [dams] "typically do more environmental harm than good."
>
> Amory Lovins (2000)

Utilities would gladly use more hydropower, *if* somebody could find a few more waterfalls. Hydropower is actually limited worldwide. Again, I cite the Ehrlichs, who continue to tell us that solar energy is inadequate to supply our energy.

> "First, if all practical sites were utilized, hydropower would only generate 1 to 1.5 TW_e..."
>
> Paul & Anne Ehrlich (1991)

For comparison, we note that the US uses 3.2 TW_t of power (all types, not just electrical); the world uses (1997) 12.7 TW_t, far more than the 1.5 TW_e that the Ehrlichs say that we can get worldwide. In plain English, hydropower could not run the world, even if *all* practical sites were utilized.

---

[44] http://www.insidechina.com/special/damkey.php3

Exercising Your Quads

Never having come to grips with the international system of units, the US government keeps track of energy production and consumption in quadrillion-BTU, known as *quads*. (One quad = 1.055 exajoules = 1.055 EJ.) Doing this disguises an interesting habit that most people might find amusing.

In 1998, says Table 1.2 "Energy Production by Source, 1949–1998," in the EIA *Annual Energy Review 1998,* the US produced 3.391 quads of energy from conventional hydropower. That's on page 7.

In the same book on page 213, Table 8.3, "Electric Utility Net Generation, 1949–1998," tells us that US utilities generated 308.8 billion kilowatt-hours from conventional hydropower. On page 215, we find that non-utilities generated 19.7 billion kWh from conventional hydropower, making a total of 328.5 billion kWh.

Let's look at the numbers. From the conversion factor above, 3.391 quads is equal to 3.578 EJ. Using one kWh = 3.6 million joules, we find that 328.5 billion kWh = 1.18 EJ. Careful study will reveal that 3.578 is not equal to 1.18.

There is nothing amiss with the conversion factors. The 328.5 billion kWh is accurate. The 3.391 quads is a fictitious, but useful, number to represent an as-if situation.

If the same 328.5 billion kWh of electrical energy had been produced by conventional (coal, oil, natural gas, nuclear, wood-fired) power plants, it would have required 3.391 quads of heat, not the 1.18 quads. The reason is that when fuels are consumed to produce electricity, there is invariably a heat engine involved, the purpose of which is to convert heat to work. The efficiency of doing so can never be 100% (see Efficiency in Heat Engines, Chapter 2). For the entire US, the average efficiency of providing electricity to the grid by burning of fuels at power plants is 33%. Therefore, when the EIA discusses hydropower, they use one figure for energy produced (328.5 billion kWh = 1.18 EJ in 1998) and another (3.391 quads = 3.578 EJ) — three times as large — for energy "consumed."

There is nothing wrong with this custom, so long as everybody knows what the EIA are doing and why. However, neither the EIA nor anybody else (such as the *Statistical Abstract*) explicitly points out what they are doing. Moreover, the use of two different units, quads and billion kWh, for what is inherently one thing — energy — obfuscates the issue.

Who's Opposed to Hydropower?

Carol Browner and Bruce Babbitt are not alone in opposing hydropower. It is axiomatic that people's lives are affected when hundreds of square miles are submerged by rising water behind dams, if only by mosquitoes. People who are personally affected naturally oppose such projects. It is also true that other people benefit from having a dam on a river. Among them are people whose lands are regularly flooded in spring rains, people who enjoy water sports, and people who use the electricity from the dam.

Eventually — in a time span that may be decades or millennia — a dam silts up. That is, the reservoir behind the dam becomes loaded with silt instead of water. The silt has no effect on the elevation head (Fig. 24), but does limit the volume of water that may be stored. So long as the reservoir can hold enough water to produce peak-capacity power when needed, there is no problem. Eventual silting has sometimes been used by environmentalists[45] as an argument against building a hydropower plant — most notably the Aswan Dam in Egypt. The fact that silting can *in principle* become a problem *eventually* is a poor argument. One needs to know, dam-by-dam, how much of a problem silting will be by when. The Mantilija Dam near Ventura, California, was evidently built in 1948 as a flood-control dam over the objections of the Army Corps of Engineers, partly on the grounds that it would soon silt up; it now has 6 billion cubic yards of sediment stored behind it. The dam is now being dismantled.

> "'The Matilija Dam,' Babbitt said then, 'is symbolic not only because it could become the largest dam to be taken down anywhere in the world, but also because it is a prime example of dams that are environmentally harmful as well as useless.'"

Texas A&M University[46]

Suffice it to say that dams are routinely opposed on environmental grounds, either because they "destroy" the landscape, block fish migration routes, or block the normal paths of land mammals.

---

[45] http://cornerhouse.icaap.org/briefings/8.html
[46] http://twri.tamu.edu/watertalk/archive/2000-Dec/Dec-28.3.html

"[Dams] seem to pose less risk of catastrophic accident than nuclear fission, although dam failures can kill thousands of people at a time—and have done so."

Paul & Anne Ehrlich (1991)

"In 1976 Leiderman ran for Congress from Missouri…primarily against the Meremec Dam project…"

Berman & O'Connor (1996)

Again we find that solar energy is opposed by environmentalists.

Chapter 9. Wind Power Is Solar Power

Heat from the sun is the source of energy for the winds, but it is not high temperature *per se* that drives the wind. For example, the Great Red Spot on Jupiter — a place not exactly well known for tropical vacations — is a cyclonic storm that has persisted for hundreds of years. Rather, it is temperature gradients that are responsible for moving the air.

In recent years, wind power has become a national fad, complete with a barrage of headlines, full-length newspaper articles, and editorials waxing eloquent about wind-generated electricity. The flyers that come with our utility bills have been encouraging us to pay extra for energy from the Clean Green Machine.

Wind Power Is Ancient. Wind-Generated Electricity is New

A picture on a piece of pottery dating from about BC 4000 shows Egyptian boats with square sails. By AD 640, wind turbines were at work in Persia grinding grain, and wind-driven gristmills have been used in Normandy since about 1180. Dutch wind turbines, used for pumping water out to the ocean, are a national symbol. Multi-bladed wind turbines have long been used on American farms for drawing water from wells. Great improvements in wind turbines followed the improvements in propeller design in World War-II. Some six thousand years after the first recorded use of wind power, we have taken the step of attaching a generator to the wind turbine to produce electricity.

Electricity is the most useful power we have, and it has been obvious for a century that wind turbines could somehow be fitted with electrical generators. The engineering problems were difficult. Most generators were designed to rotate fast, typically 3600 RPM, but such high rotation rates could only be accomplished with small-diameter wind turbines that would not be capable of producing much power.

The problem of producing electricity from the slow rotation of huge wind turbines has been addressed in two ways. Some machines have transmissions to cause rapid rotation of their generators from slow turbine rotation. Others now use modified permanent-magnet generators that rotate at the slow rate of the turbine. (Permanent magnets that can withstand the vibrations and temperature variations of an outdoor wind turbine are a comparatively recent invention.)

In the United States in 1998, wind generated 3.5 billion kWh of electrical energy, less than one tenth of one percent of the 3619.6 billion kWh generated in the country. Obviously, there has to be a reason *or reasons* that we don't get more power from the wind. Though Denis Hayes (see "Don Quixotes" page 56) blames political "roadblocks" for this situation, Tamplin (see "Puppeteers," page 44) is closer to the truth when he recognizes that economics plays a large role. But the economics all derives from the underlying physics of wind turbines. The laws of physics were not written in Washington, nor can they be repealed or flouted, even by the folks in Sacramento.

> "SMUD [Sacramento Metropolitan Utility District] continues to be a world leader in the utilization of solar technologies. The utility operates an experimental 5-megawatt wind plant in nearby Solano County."

> Berman & O'Connor (1996)

SMUD also has a photovoltaic system that produces 2 MWe at noontime on clear days if the solar cells have been recently washed. Between the two, they can produce 7 MWe if there are high winds in Solano County at the same time. That's about one percent of the power they were getting from the Rancho Seco nuclear power plant that they turned off under pressure from then-Governor Jerry Brown and other solar sirens. The year-round average power output of the two combined will be perhaps 2 MWe, so the annual output of the wind farm and the PV array is one four-hundredth of the steady *average* output of the nuke they so cavalierly discarded.

> "I would say we favor immediate phase-out of nuclear energy and a crash conversion program through alternative renewable technology."

> Jane Fonda (2000)

As they say in California: Been there. Done that.

How Wind Turbines Work

If we are to extract energy from the wind, then the wind must have energy to begin with. That energy is *kinetic energy*, the energy the wind has because the air is moving. In other words, we can extract no more energy from the wind than the wind has.

When we extract energy from the wind, we reduce the speed of the air. However, we cannot *stop* the wind entirely — extracting all of its energy —

because doing so would block incoming wind from reaching our wind turbine.

Wind Turbines, Old and New

Picture a sailing ship in a stiff wind, and you can just see the sails billowed out and you can almost feel the great force exerted on the sail by the wind. It is easy to picture why early wind turbines were built with huge canvas sails. Unfortunately, what works for sailing ships is not necessarily the best scheme for wind turbines.

High efficiency for wind turbines involves relatively thin blades that bear no resemblance to ships' sails. Modern wind turbines with two or three blades rotate more rapidly than multi-bladed farm wind turbines. Here, we are talking about angular speed, the number of rotations per minute.

Typically, the tips of the blades move about five or six times as fast as the wind. In fairly moderate wind, they move about the speed of a car on the highway. You might even wonder how they could possibly get much energy out of the wind, because there is so little surface area for the wind to hit. Perhaps surprisingly, they are much more efficient than the old sail-type wind turbines and the multi-bladed wind turbines seen on farms.

A Wind Turbine's Job Is to Stop the Air

Figure 26 is a very crude drawing of a wind turbine seen from the side, with wind blowing from left to right. If the job is to extract the kinetic energy of the wind, then the wind must be stopped! However, if we do stop the wind, then (by definition) the air in the vicinity of the wind turbine stops completely. Therefore, the main idea behind wind turbines is that they extract *some* of the energy in the wind, slowing down the air in the process.

A little thought will reveal a mild problem: how can you move high-speed air into the wind turbine when the "used-up" air leaves slowly? The answer to this dilemma is that the air leaving the wind turbine spreads out, somewhat as shown in Figure 26. The amount of air leaving the wind turbine at low velocity must equal the amount of air arriving at the wind turbine at high velocity; therefore, it spreads out into a wider area. If a tonne (1000 kg) of air comes to the wind turbine in one second, then a tonne must leave the wind turbine in one second.

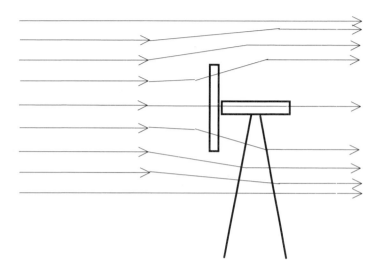

Figure 26: The wind pattern for a wind turbine.

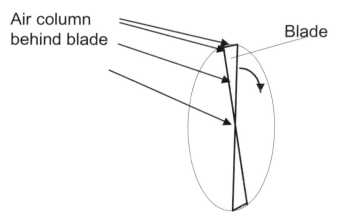

Figure 27: A thin blade blocks a column of air.

Because the air must leave with some velocity (hence some energy), only *some* of the kinetic energy is taken by the wind turbine. It turns out that only 16/27 (about 59%)[47] of the energy carried by the wind *could* be extracted by a *perfect* wind turbine; the very best real wind turbines peak out at about 50% efficiency, and only under ideal conditions.

How can skinny un-sail-like blades accomplish this task? Simply, by moving fast. At a given moment, the blade is somewhere blocking the air, and slowing down the air in a column behind the blade. But the blade is moving, and is soon blocking air elsewhere, extracting the energy from a new column of air. During the time the blade is not blocking the air in a given column, the air speed begins to increase again. However, before the wind speed has picked up significantly, the rapidly moving blade is there to take its energy again.

By this reasoning, the faster the wind turbine turns, the more efficient the wind turbine will be. However, as the wind turbine turns faster, the very blades that are extracting energy from the wind act like a fan to propel the wind, giving energy back to it. Therefore, the fan shouldn't turn *too* fast. That is, there is some ideal rotation rate for the wind turbine for any given wind speed. For multi-bladed wind turbines, the ideal rotation rate is low, and the efficiency is also very low.

Propeller Size and Windspeed

An excellent wind turbine might have nearly 50% efficiency. That is, it can convert 50% of the kinetic energy of the air that passes through it into mechanical energy. But how much kinetic energy can there be when the density of air is so low, about 1.3 kg per cubic meter, about one seven-hundredth the density of water? Obviously, it is necessary to intercept a lot of wind.

The interested reader may wish to consult Appendix B, "Equation for Wind Power" to see how wind turbine power is figured out. For a wind turbine whose diameter is D, the power output is P_{out} when the windspeed is v, and the efficiency is η. The equation is

$$P_{out} = 0.5\eta D^2 v^3 \qquad \text{Eq. 1}$$

For an excellent wind turbine (usually of two blades) the efficiency is 50%; most large three-bladed wind turbines have a maximum efficiency of about 45%. However, the efficiency η is highly variable, and deliberately

---

[47] See, for example, E. Wendell Hewson, "Electrical Energy from the Wind, pp. 6-142 – 6-174, in *Energy Technology Handbook,* Douglas M. Considine, P.E., Editor, (McGraw-Hill, 1977). The limiting efficiency is called the *Betz Limit*.

so. For example, if an ideal wind turbine is capable of, say, 2.5 MWe in a 25-m/s wind according to Eq. 1, the generator on the device may be capable of producing only 200 kW$_e$. This is always the case, because there is little point in attaching a 2.5-MW$_e$ generator that will rarely reach its maximum power. High wind speeds are pretty rare.

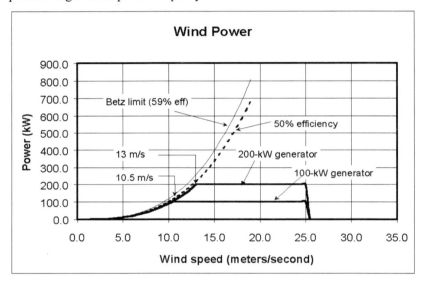

Figure 28: Wind power versus wind speed for a 20-m diameter wind turbine. The upper curve represents the power that could be obtained by a wind turbine operating at the maximum possible efficiency, 59% (the Betz limit), at all wind speeds. The second curve represents a more realistic (but still high) 50% efficiency. The third curve shows the power output for a realistic wind turbine fitted with a 200-kW$_e$ generator. The bottom curve shows the power output if the turbine has a 100-kW$_e$ generator.

Therefore, a wind turbine is fitted with a generator that is in some sense undersized. At low wind speeds, wind turbines are run so as to produce as much power as possible. However, when the windspeed exceeds some preset value, the efficiency of the wind turbine is reduced so as not to exceed the power rating of the generator.

Figure 28 shows details for a hypothetical wind turbine of 20-meter diameter for wind speeds from zero to 35 m/s. The maximum efficiency possible for a wind turbine is 59%, the so-called Betz limit. The hypothetical power from such a turbines is shown as the upper curve. The second curve shows the power from a turbine of 50% efficiency. At 35 m/s, the power would be over 4000 kW_e. Usually, wind turbines are designed to shut down completely when the speed is 25 m/s, at which time the wind power would be about 1600 kW_e. Therefore, those curves are cut off in Fig. 28 so that realistic numbers can be shown.

The lower curves in Fig. 28 are the realistic ones. If there is a 200-kW_e generator attached to the wind turbine, the power will reach 200 kW_e when the windspeed is about 13 m/s. Thereafter, the power will be held constant at 200 kW_e by varying the pitch of the blades. By the time the windspeed has reached the cutoff speed of 25 m/s, the efficiency is already down to about 6%.

With the same wind turbine, a 100-kW_e generator would reach nameplate power when the windspeed reaches 10.5 m/s, only 26% lower than the speed for the 200-kW_e generator. By the time the windspeed has reached the cutoff speed of 25 m/s, the efficiency is already down to about 3%.

Nameplate Power and Capacity Factor

Depending upon where we put our hypothetical 20-m diameter wind turbine, the annual output might be, say, 350,000 kWh with the 200-kW_e generator attached, but might be 310,000 kWh with the 100-kW_e generator. That is, the smaller generator gives somewhat less energy over the course of a year, but the power output can be held constant over a slightly wider range of windspeed. "Wind-site Predictions" on page 194 in Appendix B gives a more thorough description of the relationship between wind speed, turbine size, generator capacity, and capacity factor.

In the heady days of California's wind farms, the emphasis was on high nameplate power. They would have used the 200-kW_e generator. But then the annual energy would be a mere 20% of what a 200-kW_e unit would produce if it operated full-time at maximum power (see Appendix B). In other words, its capacity factor would be 20%.

Nowadays, the trend is to go for steadier power and a higher capacity factor. That is, one would now use the 100-kW_e generator. The 310,000 kWh produced in a year would be about 35% of the 877,000 kWh that would be possible with a 100-kW_e unit producing full power every second of the year. That is, its capacity factor would be 35%.

Just for fun, it is interesting to speculate on the capacity factor of the same wind turbine with a 2.5 MWe generator. The annual energy might be a bit higher than for the 200-kW$_e$ unit, possibly 400,000 kWh. The power factor in this case would be around 0.2%.

As we have seen, it is important to know *both* the nameplate capacity and the capacity factor for wind turbines. Just don't ask anybody in the press. They can only quote misleading garbage like "this windmill farm will produce enough power for 25,000 homes."

Case Study 1: Lake Benton I, Minnesota

Lake Benton Power Partners, LLC, operates a wind farm at Lake Benton, Minnesota, consisting of one hundred forty-three 750-kW$_e$ (nameplate) Enron wind turbines for a capacity of 107 MWe. The annual generation is 327,000 megawatt-hours. This corresponds to a capacity factor of 34.9%.

The Enron wind turbines have a radius of 23 meters. According to Eq. 1, they should produce 750 kW$_e$ at a windspeed of 11.2 meters/second, which is about 25 miles per hour. The wind turbines are designed to withstand ferocious winds of 59.5 meters per second, equivalent to 133 miles per hour. However, the machines shut down when the windspeed reaches about 25 m/s. A hypothetical wind turbine of 23-meter radius operating at 50% efficiency could produce 8300 kW$_e$, well in excess of 750 kW$_e$. However, it would be subject to extreme forces. The wind turbines are designed to produce the 750 kW$_e$ at all speeds between 11.2 m/s and 25 m/s. At the higher wind speeds, the pitch of the blades is adjusted to reduce the efficiency of the turbine, so as not to exceed the nameplate power of the generator. By 25 m/s, the efficiency is down to about 4.5%. Little is lost, however, as the wind speed is rarely that high.

We will look at six other case studies later on.

Caveat Emptor: Availability Factor

The term *availability* takes on a different meaning for wind power than it does for other sources of power. For most power sources the *availability factor* represents the fraction of time that the power is available when you need it.

For example, suppose that a hydropower station is ready at all times to produce power when it is needed. Then its *availability factor* would be 100%. As we saw in Chapter 8, the average capacity factor of hydropower stations in the US is 29.1%, but that factor comes entirely from patterns of usage. (They are designed to produce lots of power for short times.)

For another example, in a nuclear power plant, the plant is running at full power all of the time that it runs at all, typically 80% of the time. In that case, the availability factor and the capacity factor are the same, both 80%. (Recently, both figures have been about 90% because of numerous improvements.)

Power stations used at times of peak load only may operate, say, 20% of the hours during the year, but could possibly be ready at all times to turn on immediately. For this station, the capacity factor would be 20%, but the availability factor would be 100%. The power is there when you need it.

For wind farms, the term *availability factor* means the fraction of time the machine is ready to run *if the wind blows*. Wind turbines, by this definition, usually have high availability factors. As an extreme example, a mechanically and electrically perfect wind turbine sitting in an always-windless location would have an availability factor of 100% even though it never produced any electricity at all. If *availability* were defined in the same way for wind turbines as for other power stations, the figure would usually equal the capacity factor.

Constraints

In short, the wind turbine can extract *some* of the energy from the air that intercepts the wind turbine's blades. The obvious facts are:
>   the larger the wind turbine, the more power it can generate;
>   the faster the wind, the more power is available;
>   that wind turbines can't be placed one behind the other;
>   that the side-to-side spacing of wind turbines should accommodate the spread-out of air behind the wind turbines;
>   that the wind turbines should be high off the ground;
>   and that wind turbines must accommodate to varying wind speed.

Just as true are the not-so-obvious facts:
>   there is always some wind speed above which the wind turbine must be stopped, lest it fly apart;
>   that the power output of our hypothetical wind turbine is *highly* variable with wind speed;
>   that the most efficient wind turbines have the *fewest* blades;
>   that very large wind turbines introduce annoying ground vibrations,
>   and that the power output from the huge wind farms in California is a tiny fraction of California's needs.

Placement of the wind turbine

Proud citizens often like to point to a large wind turbine up on a hill and tell their visitors that the city gets a lot of their power from that lovely apparatus. Others may object that the machine is ugly. We do not intend to concern ourselves with aesthetics: if the wind turbine produces necessary power, then beauty is a secondary consideration. Do we complain about the looks of the cows that give us milk?

Far more important is the amount of wind at the site. In very approximate terms, if one site has average winds of 5 meters per second and another has average winds of 8 meters per second, the latter site will produce 4 times as much electrical power.[48] Lest there be any doubt about this matter, a research wind turbine at the University of Massachusetts was recently moved from a convenient hill near the campus to a higher hill some twenty miles away for precisely that reason.

Wind sites are classified according to the mean annual power density of the winds. "Areas designated 'good' are roughly equivalent to an estimated mean annual power, at 10 meter height, of 200 to 300 Watts per square meter (W/m^2), and 'excellent' if more than 300 W/m^2," according to a California Energy Commission report.[49] However, the average power that might be produced there would depend upon the design of the wind turbines, as we saw in Fig. 28 and associated discussion.

Arrangement for Wind Farms

It is important to understand that wind power-density figures do not refer to land area, but instead to the cross-sectional area intercepted by the spinning wind turbine's blades. Wind turbines cannot be lined up one directly behind the other because the slow air emerging from one wind turbine is of no use to the one behind. Moreover, the turbulence of wind turbine can destroy the one behind. Good engineering practice requires that rows of wind turbines be about ten diameters apart. That is, if each wind turbine has a 20-meter diameter, for example, the rows of wind turbines have to be about 200 meters apart.

Moreover, wind turbines cannot be lined up tip-to-tip, both for reasons of wind-slowing (See Fig. 26) and turbulence. Here, engineering practice is determined by the wind direction. If there is a prevailing wind direction,

---

[48] See Appendix B, "Wind-site Predictions."

[49] A. Miller and R. Simon, *Wind Power Potential in California,* San Jose State University, prepared for the California Energy Commission, May 1978.

then the wind turbines can be placed about three diameters apart, 60 meters for our wind turbine of 20-meter diameter. That is our wind turbine would sit in the middle of an area whose size is 200 meters by 60 meters, or 12,000 square meters.

What happens if we use larger wind turbines, say, 40 meters in diameter instead of 20 meters? According to the wind turbine formula (Eq. 1) we would get four times as much power at any given windspeed. But the distance between rows of wind turbines would have to be 400 meters instead of 200, and the tip-to-tip distance would have to be 120 meters. That is, the 40-m diameter wind turbine would sit in the middle of an area of 400 meters by 120 meters, or 48,000 square meters. These larger wind turbines have four times the power of the smaller ones, but the required land area is also four times as big. The result is that wind turbines can get just so much power out of a given land area.

The arithmetic is not hard to work out. If the average power density in the wind is 300 watts per square meter facing the wind, one may expect to extract about 4 watts per square meter of land area. That is, for a site with wind turbines lined up in rows that are perpendicular to the prevailing winds, you may simply divide the average power density of the winds by 75 to get the average power per unit land area.

The arrangement of wind turbines is not the same when there is no prevailing wind direction. If the winds might come from the south one day and from the east the next, then the wind turbines that were beside on one day have become ahead and behind on the next. For this case, the wind turbines have to be ten diameters apart in all directions.

For our example of 20-meter diameter wind turbines, the wind turbines should be 200 meters apart in both directions. For this case, 300 watts per square meter facing the wind results in about 1.2 watts per square meter of land area. The rule of thumb we have developed is to take the average wind power density and divide by 250 to get the power per unit land area.

The Environmental Protection Agency[50] is in agreement with these figures. It says:

> "Contemporary wind projects are typically rated at 25 to 100 MW[e]. A 25 MW[e] project might have 60 to 70 turbines covering 1500 acres. [This amounts to 16,700 watts per acre, or 4.1 W/m^2, and refers to capacity, not average output.]. Turbines, while reminiscent of aircraft propellers, are specifically designed for electrical generation.

---

[50] http://www.epa.gov/globalwarming/publications/actions/state/wa/mitigatef.html

The blade and generator housing, or nacelle, pivots to face directly into the wind. Each turbine is rated at about 300 to 500 kW[e] of *capacity.* [60 × 400 kW[e] (capacity) = 24 MW[e] (capacity)] Turbines are usually arranged in rows oriented at right angles to the direction of the prevailing winds and spaced at two to five rotor diameters from each other. Rows of turbines are usually located with roughly 10 rotor diameters between them." Wind projects typically produce power about 95 percent of the time at an average output of 28 - 35 percent of rated capacity." [emphasis and brackets added]

At a 30% capacity factor, the EPA's estimate works out to 1.23 W/m^2 of land area.

The EPA comment above is somewhat old. There are now large wind turbines that are capable of a few megawatts. Typically, they have a hub height of 50 to 100 meters and diameters of 50 to 80 meters. But the average power per unit of land area remains the same.

Case Study 2: Navitas/Minnesota

Navitas Energy, LLC, has received permission to erect eighty-seven 1.5-MWe wind turbines, for a total nameplate power of 130.5 MWe on a parcel of land consisting of 10,000 acres in Murray and Pipestone Counties, Minnesota. This translates to a peak power density of 3.22 watts per square meter of land area. With a capacity factor of 35%, this would amount to an average of 1.13 watts per square meter of land area.

Case Study 3: Chanarambie/Minnesota

Chanarambie Power Partners, LLC, has received permission to erect sixty-one 1.5-MWe wind turbines, for a total nameplate power of 91.5 MWe on a parcel of land consisting of 6,500 acres in Murray Counties, Minnesota. This translates to a peak power density of 3.48 watts per square meter of land area. With a capacity factor of 35%, this would amount to an average of 1.22 watts per square meter of land area.

Case Study 4: Big Spring, Texas

Big Spring, Texas, is home to a wind farm of 42 Vestas V47 wind turbines rated at 660 kW$_e$, and four Vestas V66 wind turbines rated at 1.65 MW$_e$. They are spread out in three groups on 23 square km (9 square miles) of land on a mesa. The nameplate energy intensity is therefore about 1.5 W/m^2. The capacity factor for the installation is 34%, so the average power production amounts to 0.51 W/m². Lightning is responsible for 4% of the downtime. Hail damage can be reduced by stopping the rotation during a hailstorm. The tips of the blades normally move at 70 m/s; adding that to the 55 m/s hail speed would increase the velocity of the hail relative to the blade to 125 m/s. The high velocity hail could cause severe damage to the blades.

Case Study 5: Kotzebue, Alaska

Kotzebue, Alaska, is just north of the Arctic circle, and is accessible only by air and water. The town has 3000 residents.

A wind farm outside of town has ten Atlantic Orient Corporation 66-kW$_e$ turbines sitting on a 148-acre site. The nameplate energy intensity is therefore 1.1 W/m². In service, the site produced an average of 83.6 kW$_e$, equivalent to an average power density of 0.14 W/m². In this installation, the wind turbines are spread out rather widely. The capacity factor for the installation is 12.6%.

The weather in Kotzebue is very dry, and icing of the propellers has not been a problem.

Case Study 6: Green Mountain, Vermont

The Green Mountain wind project is small even by wind farm standards. Its capacity is 6.05 MWe, and it has a 24.6% capacity factor. In other words, its year-round average power is 1.5 MWe. As the eleven 550-kW$_e$ wind turbines are more or less single-file along a 1.2-km line on a mountain ridge, it makes little sense to speak of the land area. The *linear* power density is 5 kW$_e$/meter. That figure might be useful for other ridge applications.

Vermont presents special problems owing largely to weather. Over 25% of the downtime was due to lightning damage. In one case, it destroyed one blade of a wind turbine. In fact, there was lightning damage about three times per year. There are about 1700 unscheduled maintenance events per year. Bad weather in the winter delayed the repairs to two wind turbines by several months. Icing has a severe impact on performance, but only for short times.

Case Study 7: South Dakota

There is a plan to put a 3000-MWe (nameplate) wind farm in South Dakota on 350 square miles of land. The nameplate energy intensity is 3.3 W/m². If they are lucky enough to have a 35% capacity factor, they will produce electrical energy at the rate of 1.16 W/m².

Scaling It Up

From the analysis and cases above, it is evident that wind farms can generate electrical power at the rate of about 1.2 W/m^2 for most sites and up to about 4 W/m^2 in the rare sites where the wind always comes from one direction. (I have been unable to find any wind farms that approach this theoretical figure.)

Now, suppose the goal is to generate enough energy to average 1000 MWe around the clock, the power output of one typical traditional power plant. At 1.2 W/m^2, the land-area requirement is about 833 square kilometers, or 300 square miles.

Let us put that land area into perspective. Imagine a 1.6-km (one-mile) wide swath of wind turbines extending from San Francisco to Los Angeles. That land area is what would be required to produce as much power around the clock as one large coal, natural gas, or nuclear power station that normally occupies about one square kilometer of land.

But the story does not end there, because California uses an average of 30.3 GW$_e$. To produce that much power would require 7,500 square kilometers (3000 square miles) of wind turbines in good-to-excellent sites with a prevailing wind direction or 25,000 square kilometers (10,000 square miles) of land area with strong winds but no prevailing direction. For comparison, that is twice the land area of Connecticut. If there were enough wind throughout the area, the 30.3 GW$_e$ could be generated with a 48-km (29-mile) swath of wind turbines between San Francisco and Los Angeles.

The scenario for getting an average of 30.3 MW$_e$ for California is quite unrealistic for other reasons. There would be times when every wind turbine produced full power, which is approximately three times the average power. In other words, they could produce an excess of 70 MWe; California couldn't use that power, nor could the power be shipped terribly far to, say, Chicago. The grid was not designed to carry the load, and isn't up to the task. Therefore, the output of the wind turbines would have to be cut back by 2/3. The overall capacity factor would suffer. Therein lies a reason that wind turbines can't provide a huge percentage of the average electrical power on the grid.

Altamont Pass in California is one particularly good site for wind turbines, at least according to my experience there in July, 2001 when the wind was howling. Clearly, the wind power density there is much higher than it is over the billions of square meters (thousands of square miles) that would be required to produce a significant fraction of California's electricity.

Figure 29: A few of the wind turbines at Altamont Pass in California.

It is true, of course, that wind turbines have a relatively small footprint. That is, although a 100-MWe (average power) wind farm might occupy 100 square kilometers, the windmill towers themselves cover only, say, 1 square kilometer. The land can be used for farming or ranching at the same time.

Nevertheless, the visual effect is stunning. Make no mistake about it, even a small wind farm is an impressive sight. Figure 29 shows a few of the hundreds of wind turbines at Altamont Pass, elevation 300 meters (1000 feet), east of Livermore, California. The site has no trees, and receives strong ocean breezes. It is apparent to the casual visitor that the machinery is huge and hazardous.

There are times when the wind is calm *everywhere*. At those times, all of the power would have to come from somewhere else. Engineers are fond of saying that wind turbines "provide energy but not power." It's jargon, to be sure, but they have a point. Wind turbines do most assuredly reduce the use of fuel; however, they do not allow a utility to get rid of so much as one power plant. The utilities must maintain full reserve to handle the situation when the wind does not blow. In other words, wind turbines do not add meaningful capacity to a system.

To put it fairly, but bluntly, wind power can never produce more than a small fraction of our electricity.

"At home it [Denmark] has 2000 grid-linked turbines on line, with a capacity of 250 MW[e]..."

Jon Naar (1990)

"But, like hydropower, wind power can make only a modest contribution to save us from the energy trap. If the windiest 3 percent of Earth's land area were covered with efficient wind turbines (a gigantic undertaking), only about 1 TW[e] of electricity might be generated..."

Paul & Anne Ehrlich (1991)

Small Units for Home and Farm

Wind turbines are probably the cheapest source of energy for most remote applications. The utility is perfectly happy to run power lines to a customer who will be using a megawatt of power, but it's awfully expensive to run power lines 50 kilometers to a farm or home that will use no more than a few kilowatts.

A home that gets its electricity from a wind turbine needs deep-discharge batteries to supply power when the wind isn't blowing, and a big-enough wind turbine to recharge the batteries and run the household when the winds are up. The installation is usually expensive, but nowhere near the cost of running long power lines from the utility. I know of one mountain household near Westcliffe, Colorado, is not connected to the grid at all. The owners use a combination of wind and photovoltaics. They have lots of big batteries to store electrical energy. They also have to be prudent in their consumption of power. They are proud to be able to survive without the electricity from the grid, but would tie into the grid in a heartbeat if they had the opportunity.

Figure 30 shows the performance of a commercial 1-kW$_e$ wind turbine that uses a modified automotive alternator as its electrical generator. It produces about 1 kilowatt between wind speeds of 11 m/s and 20 m/s. Just like its larger cousins, this unit is designed to have relatively constant output at wind speeds above about 10 m/s. At the lower wind speeds, however, the efficiency is about half that of large commercial units.

Wind Power

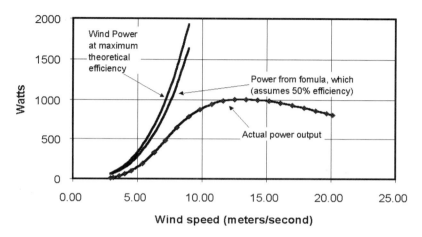

Figure 30: The performance of a commercial 1-kWₑ wind turbine.

A Few More Problems

The Wind Turbine Meets the Power Line

The wind rotates the turbine and the generator produces electricity. How is the turbine's electricity to be matched to that of the utility? Imagine that you have two wires, one in each hand. One wire has electricity provided by the utility and the other wire's electricity comes from a wind turbine (or, for that matter, from any other source). Your job is to connect the wires so that there is no spark.

If — at all times — the voltage on the wire in your left hand is identical to the voltage on the wire in your right hand, the wires may be connected together without any sparks. The voltage on a 115-VAC power line varies from +163 volts[51] to −163 volts in 1/120 second and back to + 163 volts

---

[51] A 115-VAC power line makes a light bulb burn just as brightly as a 115- volt battery does. Since the A.C. voltage varies from positive to negative and is low most of the time, it is clear that the voltage must exceed 115

during the next 1/120 second. (The AC voltage has the same heating power as 115 volts DC.)

The wind turbine's voltage, to be properly tied to the power line, should have exactly the same voltage, exactly the same frequency, and exactly the same timing in order to match the power line voltage. In the case of large turbines, the voltage to be matched is the high voltage of the transmission line, but the principles are the same — the output voltage of the generator must match the power-line voltage at all times.

Three techniques have been used to provide the match. In one, induction motors have been run as generators, with the AC power line providing a rotating magnetic field in the generator. The RPM is controlled by the pitch of the blades. In a related system, variable "slip" (see next paragraphs) also uses the AC power, but introduces variable slip to allow for variations in RPM. Finally, DC generators (often using permanent magnets) can generate DC, which can be converted electronically to AC to match the power line.

Pardon, your slip is showing.

In a typical induction motor such as the one in a washing machine, the line voltage causes a magnetic field to rotate at 3600 rotations per minute (60 cycles per second), but the rotor rotates at 1875 rotations per minute. The difference, 1725 revolutions per minute, is called the *slip RPM*, and the fraction (3600 − 1875)/3600 = 0.48, or 48%, is called the *slip*.

In an induction *generator*, somewhat the same principles apply, but the applied torque (from a steam generator, for example) causes the rotor to rotate *faster* that 3600 RPM. That is, the slip is present, but of opposite sign. The rotor is ahead, not behind.

In wind-driven induction systems, multi-pole generators (see Fig. 31) do not need to turn at 3600 RPM to generate 60-Hz electricity. For example, an 8-pole generator needs to rotate at only 900 RPM. Still, that is very fast compared to the rotation rate of the turbine itself, which may be in the range of 20 to 100 RPM (depending upon turbine diameter). Transmissions are used to convert the low RPM of the turbine to the higher RPM required by the generator. Still, the generator operates on the slip principle — the rotor turns faster than the magnetic field, always adding energy to the power line.

---

volts considerably for some of the time. In fact, the voltage rises to $115\sqrt{2}$ = 163 volts, and goes negative by that amount as well.

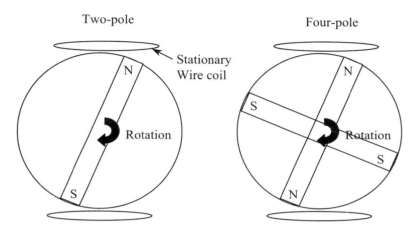

Figure 31: The rotation of an electromagnet in a motor or generator. The current-carrying coils that produce the magnetic field are not shown. On the right is a four-pole magnet. In a generator, the voltage is induced in the stationary wire coils. The slower is the rotation rate, the greater is the number of poles required to produce the desired voltage.

In a simple induction system, the kind of system used by most early wind turbines, the turbine always had to turn at some fixed RPM. The RPM adjustment would be done by varying the pitch of the blades. That is, if the RPM were to increase, the frequency of the generated electricity would increase. In more modern induction-generator systems, it is possible to vary the RPM of the turbine by varying the amount of slip. An optical signal is sent to an electronic circuit on the rotating shaft that causes it to adjust the slip to compensate for varying wind speed. It is actually easier to adjust the slip electronically than to adjust the turbine's RPM mechanically.

Whether the induction generator is one of constant slip or optically controlled variable slip, the power line itself is used to produce the magnetic field. By itself, the generator is incapable of generating voltage. Moreover, since it is an alternating-current system, the voltage must be continuously applied.[52] In Hawaii, when the winds were strong, the utility had to fire up a diesel generator in order to provide the current for the magnetic fields of

---

[52] In engineering jargon, the induction motors "suck up VARS" (volt-ampere, reactive).

the wind farm. After the wind farm was required to pay for that service, they were no longer able to stay in business.

Words from the PM

Although general physics books often describe generators in terms of permanent magnets that move past coils of wire, there are many practical reasons that such generators have not been used. Recent developments, however, have made permanent-magnet generators practical. Especially, we can make large, strong, durable magnets that can maintain their magnetic fields despite physical vibrations and years of use.

Usually, a gear mechanism is necessary to convert the slow RPM of the turbine to a high RPM necessary for an ordinary generator to work well. The array of magnets around the periphery of the rotor in Fig. 32, however, allows for the generator to be driven directly. Stationary coils experience a rapidly changing magnetic field, which induces a voltage. The faster the change, the higher the voltage. In most generators, there is either one north-south pair or two north-south pairs of magnetic poles, and they are usually electromagnets. The large array of permanent magnets in Fig. 32 allows for a low-RPM rotation of the rotor to move the magnets past the coils rapidly even at low RPM. Therefore, the rotor can spin at the same lumbering rate as the wind turbine and no gearing is necessary.

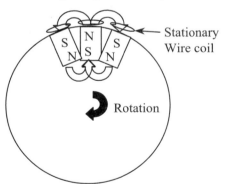

Figure 32: A permanent-magnet generator with numerous permanent magnets around the periphery of a rotor. The rapidly changing magnetic field through each stationary coil induces a voltage.

In such systems, the usual practice is to generate DC voltage and then use an inverter to match the voltage to that of the power line.

Maintenance

Wind turbines are high-maintenance machines. On a recent trip to Altamont Pass, I saw that most of the wind turbines were spinning; however there were numerous exceptions. Perhaps 80% to 90% of the horizontal axis machines were at work, but the vertical-axis turbines (Fig. 33) were overwhelmingly *not* spinning. Possibly it is for this reason that Amory Lovins, Flavin *et. al.*, and Jane Fonda argue that solar energy creates jobs.

Figure 33: Vertical-axis wind turbines at Altamont Pass. In July, 2001, under windy conditions, most of them were *not* rotating.

One problem that besets wind turbines in some locations is the buildup of ice on the blades. Of course, that happens only where there are ice storms, namely, most of the US. It doesn't take much of an imbalance to cause violent shaking of the entire structure. Automatic controls can presumably shut the wind turbine down under such conditions. At the present time, according to *Photonics Tech Briefs* magazine, NASA, MIT, and Visidine are engineering an infrared sensor to detect icing on helicopter blades, which suffer the same kinds of problems as do wind turbine blades.

Yet another problem, for some places at some times of the year, is the presence of dead bugs on the turbine blades. High efficiency requires that the blades be smooth, but the dead bugs roughen the edges[53]. European scientists have discovered that a bug buildup of a millimeter or two on the leading edges of turbine blades can reduce power by as much as 25%.

Safety

It is often advertised that wind turbines are safe, and that they leave most of the land available for crop production. Figure 34 shows one of several collapsed towers I saw from the Altamont Pass Road in July of 2001. I don't imagine that Altamont Pass was a pleasant place to be on the day(s) that the towers keeled over. Nor would I like to be an insurance agent for a wind farm that has kids running around getting into mischief.

There is probably more hazard associated with numerous small wind turbines than with large wind farms. That is, one thousand 1-kW$_e$ turbines are more likely to inflict injury than a single 1-MW$_e$ turbine. In a recent windstorm in Southern Colorado, several wind turbines of the 1-kW$_e$ class were destroyed by the wind[54]. These wind turbines were for home and farm use, and were therefore located close to people.

One design problem for wind turbine engineers is that any thing that rotates has the same peculiar properties as a gyroscope. For example, a wind turbine might be rotating because of a wind from the east. If the wind direction changes so that the wind comes from the southeast, then there will be an upward or downward twist (depending upon the direction of rotation) on the assembly that holds the wind turbine on the tower.

There was a recent case of a death of a workman in Hawaii due to the dreaded LOBA — Loss-Of-Blade Accident.[55] Whether the LOBA was due to the gyroscope effect or to some other cause, the point remains that the workman was killed. In some locations raptors, including American Bald Eagles, have been killed in large numbers by wind turbines operating normally.

---

[53] Yahoo News, 7/04/2001

[54] Information provided by a local vendor of wind turbines.

[55] Alan S. Lloyd, Professional Engineer, private communication, 11/13/01.

Figure 34: Collapsed wind-turbine at Altamont Pass.

The specifications for a certain 600-kW$_e$ wind turbine on the market are worthy of study. It has a diameter of 44 meters (144 feet), and produces its rated output at 28 RPM. At that rate of rotation, the tips of the blades are moving at 64.5 m/s, which is 145 miles per hour. No wonder some spokesperson for the Sierra Club has dubbed them *Cuisinarts of the Air*.[56]

Transmission Lines

The generators themselves produce voltage at somewhere between 480 volts and 4800 volts, and a transformer at the base of the tower raises the voltage to either 12.47 kV or 24.9 kV (in Colorado, at least). The transmission lines that carry the power are underground to protect them from the turbine blades. In a wind farm, the transmission lines form a grid of some layout that depends upon the number of turbines and the overall arrangement of turbines. Regardless of how it is done, some cables have to carry the current from numerous turbines.

For example, assume that ten 1-MWe generators are tied in series by one-ohm power lines. The power from a 1-MWe turbine (#1) is delivered at 25-kV by a one-ohm cable to the next turbine (#2). The power loss in that one line would be a negligible 1600 watts. The transmission line between turbines #2 and #3 would be 6400 watts. Between #3 and #4, the power loss would be 14.4 kW$_e$, and so on. The one-ohm power line after #10 would have a power loss of 160,000 kW$_e$. The power losses add up to about 600,000 watts, which is 6% of the power generated. But a 1000-MW$_e$ wind farm would require 1000 1-MW$_e$ generators, not a mere 10.

This situation is considerably different from that of a large power station that generates 1000 MW$_e$ in one location. No matter what the layout of wind turbines, the problem of keeping power losses within reason amounts to using the highest voltages possible, consistent with underground power lines, and laying out a grid work that keeps wires from becoming overloaded.

Aesthetics

We are accustomed to seeing high-voltage transmission lines, and we are all aware that citizens object to them on some occasions, largely for reasons of aesthetics. Wind farms have also been opposed — in Denmark, England, Australia, and Minnesota, to name a few places — on grounds of aesthetics.

The large wind turbines — 500 kW$_e$ and up — dwarf the towers used for high-voltage transmission. But unlike transmission lines that cut a swath that extends for miles, wind farms cover wide areas, miles in *both* directions.

---

[56] Michael Fumento, "Good News, Bad News," *Reason*, (June, 2000). (The Sierra Club has recanted the description.)

In some places, pressure from environmentalists has forced utilities to run their transmission lines underground, at horrendous expense. The aesthetic arguments against transmission lines apply even more strongly to wind turbines. I can hardly wait to hear somebody demand that wind turbines be placed underground.

Power Quality

The electricity produced by steam turbines, whether the energy that produces the steam comes from coal, oil, natural gas, or nuclear fission, is of very high quality, because the rotation rate of the generators is nearly constant. Generators turned by diesel engines, by contrast, rotate slightly non-uniformly because of the periodic push of the pistons. The effect can be observed in the power put onto the transmission lines, showing up as overtones of the normal 60-Hz of the power lines. In fact, since the entire grid is tuned to handle 60-Hz power, all of the power in the overtones is eventually lost to heat.

Wind power, according to utility engineers who have to work with it, is the lowest quality power on the planet. The problems are of no particular concern for the rural homeowner, but they are very burdensome to utilities that are required to deliver high-quality power to their customers. One effect is periodic pulsing in the electrical power occasioned by the passage of turbine blades through the "shadow area" of the tower; most wind turbines operate with the turbine blades on the leeward side of the tower.

The Engineers Chase Frequency

When a large load is connected to the power line, the voltage drops, but the more evident change is a drop in the rotation rate of the generators. Electronic sensors immediately detect the resulting decrease in the frequency of the AC electricity. Similarly, when a large load is disconnected from the power line, the frequency increases.

Electric clocks that are plugged into the power line rely on the frequency of the power to be exact. If the frequency — 60 cycles per second — varies by 1 part in 100,000, the clocks will be off by about a second in one day.

Sudden changes in load are not the only ways to cause fluctuations in line frequency. The same sort of thing can happen when a given power plant suddenly comes on line or goes dead.

There are control stations staffed by dispatchers that monitor the power lines. If the frequency varies ever so slightly, they must respond by calling for more (or less) power from the power stations whose power levels are easiest to vary — hydro stations, when possible, otherwise "spinning reserve".

When the wind is gusty at a wind farm, the power can fluctuate wildly, and in short order. If the power output is in the flat region of Fig. 28, the wind speed does not affect the power output. However, the capacity factor — typically about 20% to 30% — tells us that wind farms spend most of the time operating in the low-power region where the output power is very sensitive to wind speed. In a gust, the power increases, and the effect is to increase the frequency of the power line. In a lull, the frequency decreases.

The engineers get very exasperated "chasing frequency," trying to hold the line frequency constant while the wind causes the input power to fluctuate wildly. They have to call for more or less power from conventional plants to keep up with the whims of the wind. Doing so causes unnecessary wear on the conventional plants.

Dispatchers can compensate for fluctuations in wind power only when the wind farms supply no more than about 10% of the power in a huge grid. In the words of one utility engineer [private communication]:

> "Utilities have the right to control large fluctuating loads (arc furnaces for example) so their other customers aren't impacted. We should also have the right to control fluctuating sources. But, wind turbines are 'green' so we when we raise these concerns we are accused of 'impeding' renewable energy."

Again, if wind were a viable power source, utilities would be champing at the bit to use it. Utilities use every technology available to cut their fuel costs; they would gladly use photovoltaics and wind turbines *if* they produced economically.

Let me be a little more specific about that matter. All utilities know the exact cost — down to a fraction of a penny per kilowatt-hour of producing electricity from every individual power plant. When there is an increase in demand, the utility automatically adjusts things to get the cheapest electricity available. When there is a decrease in demand, they automatically turn down their most expensive power plant. Utilities generally have to be *forced* to buy wind power.

> "Nuclear power has flunked the market test."
>
> Lovins & Lovins (2001)

Nuclear power plants run full time, generating 21% of utilities' electrical energy. Presumably the technosolar systems that bring us 0.12 percent of our energy *have* passed the market test.

The EIA recognizes (*International Energy Outlook 2001*) that wind power exists because of subsidies.

Wind power in the United States enjoyed substantial growth in 1999, mostly because of the threatened expiration of Federal tax credit for wind production in June 1999 (which has since been extended to the end of 2001).

During that that subsidy-favored year of 1999, wind production in the entire US increased from 2,998 million kWh in 1998 to 4,488 million kWh. The increase in average power was therefore 170 MWe, and the average power production from all wind turbines combined in 1999 was about 511 MW_e, about the amount provided by any medium size power plant.

Environmental Considerations

The main complaints at this stage have come from bird lovers who have complained about dead birds, chiefly large raptors. Golden Eagles, for one, have been killed by the rapidly moving blades of wind turbines. Other complaints have been about noise (sound), ground vibrations (from large, low-RPM machinery), and TV interference. Some people have complained that the wind turbines are ugly. Aside from the physical hazard to raptors (and anything that gets in the way when a wind turbine flies apart), these problems could not be considered environmental.

But things have just begun. Wind provides a trifling 0.10% — one part out of a thousand — of our electricity. Were it to provide 7% of our electricity (Table 1, page 41) as the Union of Concerned Scientists predicted for 2000, there would have to be 70 wind turbines for every one in existence. (That is, of course, an underestimate, because the first sites to be used were obviously the best ones.)

> And while wind farms appear to use large tracts, only 10 percent of the land is occupied by turbine towers and service roads; ..."
>
> Brown, Flavin & Postel (1991)

The scenes at Altamont Pass (See Fig. 29, which shows only a tiny fraction of the wind farm) would be repeated endlessly. How well would wind turbines be accepted in places where they have strong ocean breezes — for example, on the Monterey Peninsula? "Come play golf at Pebble Beach and visit our beautiful wind turbines!" "Pick up some pre-diced gulls!"

Arguably, these are not serious environmental problems in and of themselves. But small environmental significance has not stopped lawsuits in the past. Nor, one expects, will it in the future.

It's a pity, perhaps, because, aside from hydro and biomass, wind is the best solar prospect.

Chapter 10. Direct Solar Heat

Tours of solar homes tell us two important things. (1) Solar heating is a possibility, and (2) some movie stars can afford them. It does not follow that most people can afford solar homes. In fact, we have seen in Fig. 11 that a mere 0.03% of all US homes — one out of 3330 homes — is heated by solar collectors.

The Greenhouse Effect

The usual explanation for the greenhouse effect is overly simplistic, especially for a solar collector that is feeding heat into a house or a hot-water system. Let us look at what happens to solar energy in a solar collector.

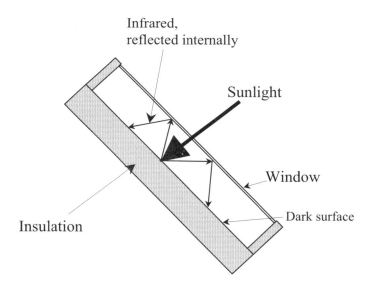

Figure 35: A solar heat collector. Sunlight passes through the glass and strikes the dark surface, which absorbs the energy and becomes hotter. The surface radiates infrared (IR) radiation, but the window blocks that radiation from leaving. The solar collector therefore retains the heat.

The design of a solar-heat collector is simplicity itself. It is merely a box with a glass (sometimes transparent plastic) window facing the sun. Figure 35 shows a typical solar heat collector, and the caption gives the standard explanation: Sunlight goes in, infrared can't get out, and the collector gets hotter.

To understand what is wrong with the standard explanation, it is only necessary to ask what would happen if a solar heat collector sat in the sunlight all day. Would it continuously get hotter? No, it would heat up to some maximum temperature and then heat up no more. For example, I assembled a crude solar collector and put in into the bright Pueblo, Colorado, sunlight in early June when the outdoor temperature was 33 °C (92 °F). The interior temperature initially rose rapidly, then more slowly, eventually leveling off at 92 °C (198 °F). Other collectors have similar behavior.

Why doesn't the collector keep getting hotter and hotter? Think of it this way. You have a room with huge windows, and you try to keep the room hot enough to boil water. Obviously, you would have to pour in a lot of heat because there is a lot of heat lost through those windows. The windows block IR, to be sure, but they do also *conduct* heat. Eventually the solar collector reaches an equilibrium temperature where the rate of heat loss through the glass and through the walls exactly equals the rate that heat comes in via sunlight.

When the temperature of the surroundings is colder, the equilibrium temperature of the interior is also lower. The reason is simple. The rate of heat loss is proportional to the temperature difference between the inside and the outside.[56] For example, the collector I tested achieved a temperature that was 41 °C (106 °F) above that of the surroundings. On a freezing day (0 °C, 32 °F), the interior temperature could rise only to 41 °C (138 °F).

Typical commercial glazed solar collectors are better than the one I slapped together out of a piece of glass, a box of insulation and a black plastic bag for a heat absorber. They can often achieve a temperature rise of about 95 °C (170 °F) above their surroundings. (I use round numbers for both Fahrenheit and Celsius.)

And, what was the efficiency of my solar collector when it reached 92 °C? A big fat zero percent. The efficiency is the ratio of the energy delivered (in this case, heat) to the energy input. No energy was delivered to anything. Zero divided by something is still zero.

If I had built the solar collector so that it could deliver heat to some target (water, or room air, perhaps), the temperature of the collector would drop somewhat, depending upon how much heat it delivered. By vigorously moving water through the system, I could achieve an efficiency

---

[56] See Insulation and *R*-value in Appendix B, page 191.

of about 80 percent. (The glass reflects sunlight to that extent.) That is, if the collector were not allowed to warm up above the temperature of the surroundings, the rate of heat loss would be zero. *All* of the heat collected from the sun would be delivered to the target. (We are ignoring the power used in pumping the water around.)

On a summer day when the temperature is 33 °C (92 °F), most people would be loath to heat their houses with sunlight or anything else. Home heating is something one wants on cold days. If the outdoor temperature were 0 °C, the solar collector could still be made to operate at nearly 80% efficiency. Just circulate enough water (or air) to keep the temperature of the collector at the freezing point. But that's a pretty poor way to heat a house — using freezing air. (I'm told it's an old trick that slumlords use.)

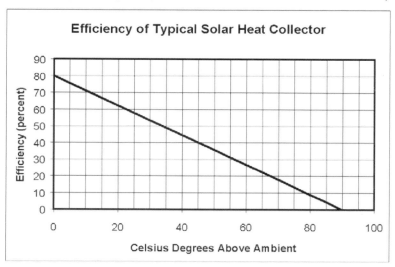

Figure 36: The efficiency of a typical commercial solar heat collector, versus the difference between internal and external temperatures. Solar intensity of 950 W/m² is assumed.

To heat a house usually requires the temperature of the circulating air to be at least 50 °C (120 °F). That is well above the ambient temperature, so it is obvious that the collector will lose heat to the surroundings. Moreover, the temperature of the collector will be above the temperature of the moving air, typically 5 °C or more, meaning that the collector itself would be 55 °C.

Therefore the efficiency will be well short of the maximum 80%. We will use the data from Fig. 36 for a typical commercial collector to find the efficiency under the circumstances given: solar intensity of 950 W/m², 0 °C ambient temperature, 55 °C interior temperature. The efficiency would be 31%.

If the ambient temperature were 10 °C lower, namely –10 °C (14 °F), the collector would still have to be 55 °C, therefore 65 °C above ambient. The efficiency would be 22%. Similarly, if the ambient temperature were 10 °C (50 °F), the efficiency would be 40%. The colder the ambient temperature, the lower the efficiency of the collector. Usually, of course, the colder the ambient temperature, the lower the quantity of available sunlight. This is exactly the opposite of what one would desire.

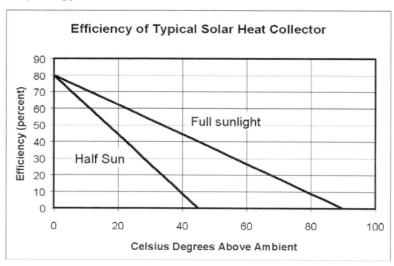

Figure 37: The efficiency of a typical commercial solar heat collector, versus the difference between internal and external temperatures, shown for solar intensities of 950 W/m² (Full), and 425 W/m² (Half).

The intensity of sunlight is never 950 W/m² except in low latitudes when the sky is clear in the summertime. When the sun is blocked by haze, the solar intensity can easily drop by half to 425 W/m². Under those circumstances, the efficiency curve looks a bit different. The interior temperature can never rise 90° C above ambient; in fact it will rise at most 45 °C above ambient.

Figure 37 shows the efficiency of the same collector for a solar intensity of 425 W/m². At 0 °C ambient temperature, the interior does not rise enough to provide meaningful heat to the house. That is, the efficiency is zero.

The half-sun condition does not need to come from cloudy weather. The angle of the sun's rays depend upon the latitude and the time of day, and that angle has an effect on the intensity. When the sun's rays are 60° from being perpendicular to the surface, the solar intensity is effectively halved.

Avoiding the Expense

On page 59, we reviewed the well-known rules for home heating, of which we repeat the first.

Rule #1: Insulate your house so well that *even* solar energy can heat it.

Sunday supplements extolling the features of solar homes never fail to explain how the builder installed gobs of insulation. Unfortunately, they always talk about the home as if it were a *solar* home. Invariably, it is a well-insulated home with a standard heating system. The solar collector system is often an expensive add-on that is of little consequence. It is the *insulation*, not the solar collectors, that deserves the credit for decreasing heating bills.

Solar collectors are expensive and usually unsightly. Large windows are expensive and often beautiful. Therefore, the preferred way to use sunlight to heat a home is to install large south-facing windows with a roof overhang that allows winter sunlight in and blocks overhead summer sunlight. This so-called *passive-solar* heating is not included in the EIA's count of solar homes. In fact, probably most of the homes in the US can lay claim to *some* solar heat, if only because sunlight enters windows some time during the day. The figure given in the caption of Fig. 11 that a mere 0.03% of homes is solar-heated refers only to those homes that have external solar collectors.

Some homes built according to this model have large masses of stone (or water) that warm up in the sunlight and release heat at night. They have the double virtue of storing heat for the night and keeping the house from getting *too* hot during the day.

The disadvantage of passive solar is that windows are not nearly as good at retaining heat as insulated walls are. Therefore, it is necessary to install heavy curtains or other insulation to keep heat from escaping at night.

Heating the Pool

Heating swimming pools in the summer is an entirely different matter from heating homes in the winter. Plenty of sunlight is available, the storage system is the water itself, and the collector never needs to be hot.

Imagine a serpentine array of black PVC pipe on a south-facing roof as shown in Fig. 38. There is no glass to reflect any light, so the efficiency is not limited to 80% as it is with typical glazed collectors. Figure 39 shows the performance of a commercial solar heat collector for a swimming pool. The light line is for a typical solar collector for home heating. The heavy line is for a commercial solar heater for swimming pools. When the temperature of the collector is not far above that of the surroundings, the efficiency of the collector can approach 90%.

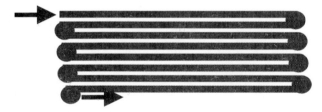

Figure 38: A primitive solar swimming pool heater. There is a serpentine array of black plastic pipe on a south-facing roof. The water from the swimming pool is pumped through the system, heating the water and cooling the collector. The collector never gets hot enough to lose much heat to the surroundings; therefore the efficiency is very high.

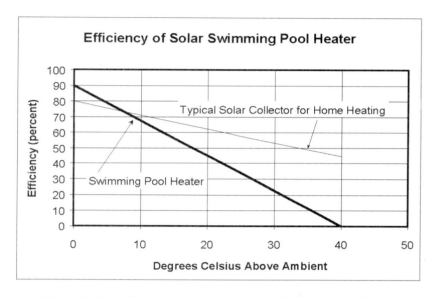

Figure 39: The efficiency curve for a commercial swimming pool solar heat collector (heavy line) compared to typical solar collector for home heating (light line). When the pool heater is only slightly warmer than its surroundings, its efficiency approaches 90 percent.

The solar collector for heating swimming pools thus satisfies several important criteria: low cost, high efficiency, and simplicity. It wins no points in a beauty contest.

In arid climates, the most important mechanism that cools swimming pools is evaporation. Simply covering the pool with a clear plastic sheet has the dual advantage of limiting evaporation and letting sunlight in to heat the pool.

Domestic Hot Water

Using sunlight to heat domestic water up to the temperature required for a dishwasher is necessarily inefficient, just as Fig. 36 suggests. However, water emerging from pipes in the ground (typically at 10 °C, 50 °F) can be pre-heated efficiently by a solar collector to, say, 38 °C (100 °F), especially in the warmer months. The regular water heater can then finish the job.

There are books that tell how to make and install domestic solar water pre-heaters of several designs, and I will not attempt to steal their thunder.

"Kazimir's prize-winning Pacemaker water heater, designed for a household of two, uses existing water tanks, sells for about $700, and can be installed by a non-plumber who is handy with tools. Though it is on sale at several Home Depot outlets, the product has sold better in the Caribbean than in Florida."

Berman & O'Connor (1996)

No doubt the sales of the Kazimir are better in Florida than in Minnesota. Curiously, solar swimming-pool heaters are a poor seller in Hawaii. The reason is that heat pumps consume less electricity than the water pumps necessary to use solar energy for the same job.[57]

The Market for Solar Heat Collectors

The US Energy Information Agency (EIA)'s *Annual Energy Review,* both in book form and on the web at www.eia.doe.gov, gives data on the shipments of solar collectors. During the last several years, about 680,000 square meters of solar heat collectors were made and shipped with the US, or imported to the US. Another 40,000 square meters were exported.

If all of the solar collectors sold in the US from 1979 to 1998 are still in service, the total is 20.9 million square meters, equivalent to 8 square miles). Figure 40 shows the annual shipments (including exports) of solar thermal collectors from 1987 till 1998. (The graph is taken from the EIA website, which still uses the British system of units.) High-temperature thermal units (the right-hand bars in Fig. 40) went off the market after 1990, and the predominant type of solar collector has been the low-temperature type. The low-temperature collectors are, after all, the most efficient ones.

Not surprisingly, the overwhelming use for solar collectors is the heating of swimming pools. Those heaters constitute 92% of the market (measured in surface area of solar collector). Solar collectors for domestic space heating and for hot water are about 0.8% and 2.9% of the total market, respectively.

---

[57] Alan S. Lloyd, Professional Engineer, private communication 11/13/01.

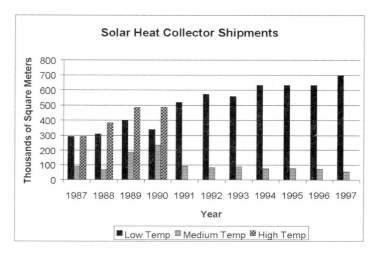

Figure 40: Shipments of solar thermal collectors, in thousands of square feet (graph in archaic units from www.eia.doe.gov). Some 93% of the collector shipments are low-temperature collectors.

It should surprise nobody that the chief buyers of solar-thermal collectors reside in just a few states. The top destinations are Florida (51%), California (23%), and Arizona (6.5%).

More Sobering Statistics

In the US, 5.4% of our energy is used for heating homes. An additional 2.0% of our energy is used for heating water in homes. Heating of commercial establishments requires 1.9% of our energy, and heating water in commercial establishments requires another 0.9%. Considered together, heating of the air and water in homes and commercial establishments accounts for 10.2% of our energy consumption. If, by some miracle, we could suddenly get *all* of our domestic and commercial space heating and hot water from sunlight, we would have solved 10.2% of the energy problem.

What about the other 89.8%?

And that 10.2% isn't going to happen anyway.

Chapter 11. Free Electricity!

Benjamin Franklin experimented with electricity, but he had no idea how it might be used. Michael Faraday, asked by a member of the British Parliament what earthly good could come of his research into electricity, replied that he did not know, but that "some day you will tax it."

Electricity is our most versatile form of energy. It can provide light, motion, and the power for all of the ubiquitous electronics that are a part of our daily lives. It is not — presently — well suited to transportation, except for commuter railways in cities like Boston.

There is something very appealing about converting sunlight directly to electricity with no moving parts. "Obviously" (it says here), there is nothing to wear out. All we need to do is to put something out in the sunlight and we'll have electricity.

There is nothing particularly new about the concept of converting sunlight to electricity. In 1822, T. S. Seebeck obtained an electric current by heating one of the junctions of a bi-metallic ring, and it was only a matter of time before thermocouples using the Seebeck effect were in use as temperature-measuring devices. Let the source of heat be sunlight, and there's solar electricity, albeit of low efficiency.

Another way to use sunlight to produce electricity is to focus the sun's rays to heat water (directly or indirectly) to a high enough temperature to run a steam engine.

Yet a third way, discovered in 1878, but commercially developed within the last few decades, is to produce electricity directly, without having to produce heat. Everybody these days is familiar with solar-powered calculators, for example. It is this new, somewhat exotic, *photovoltaic* technology that has some people very excited about solar energy for the future. In the view of some, it is an infant technology that will eventually become cheap enough to enjoy widespread use. After all, solar cells are made of silicon, the same stuff that computer chips are made of. "Look how the price has been dropping!" they say. Don't count on it.

Some Solar Installations

There are two major installations in the US devoted to producing electricity from the heat of sunlight. Of course, sunlight has to be concentrated to produce the high temperatures necessary. The installation at Solar Two near Barstow, California, shows one method to accomplish the task, and the parabolic-trough array at Daggett, California, shows another.

Both installations are in the best solar location in the US, namely, the Mojave Desert.

Solar Two

In a unit called *Solar Two*, shown in Fig. 41, computer-controlled mirrors reflect light onto a central tower where the concentrated sunlight heats a heat-transfer oil called *therminol*. The hot therminol is pumped to a heat exchanger where the heat is transferred to water to make steam to run an engine. The engine is coupled to a generator to produce electricity.

It is important to understand that magnifying glasses and mirrors *concentrate* light, but *do not create it*. We all learned, as children, how to cause high temperature by focusing the sun's rays into one bright spot.

Figure 41: Solar Two in the Mojave Desert as seen from the air. Computer-controlled mirrors direct sunlight to the tower, where the high intensity of light heats oil that in turn heats water to make steam to drive a turbine to produce electricity. Photograph courtesy of http://www.nrel.gov/data/pix/searchpix.html, taken by Joe Flores, Southern California Edison

The installation is called *Solar Two* because *Solar One* burned up in a fire on August 31, 1986 when 240,000 gallons of heat-transfer oil caught fire. The fire did not destroy (most of) the mirrors, however, because they are spread out over a large area. A new tower was constructed, and new experiments were begun.

Carol Browner, the head of the EPA in the Clinton administration, is shown in Fig. 42 giving a pep talk at Solar Two.

Figure 42: Clinton's EPA Head Administrator Carol Browner gives a pep talk to the mirrors at Solar Two. Her microphone is probably not solar powered. Photograph courtesy of http://www.nrel.gov/data-/pix/searchpix.html, taken by Warren Gretz.

There is much in common here with the problem of being able to see a movie screen in an auditorium where all of the seats are on one flat floor. Somebody's head is always in the way. If the screen is low, the only people who can see it are the ones in the front row; if the screen is raised, the view improves for the ones near the front, but remains bad for those at the rear, until the screen is very high indeed.

Figure 43 shows a two-dimensional sketch (looking south at about 9:00 a.m. in mid-summer) of how the solar mirror assembly works. Light from the sun, now somewhat low on the southeastern horizon, reflects off the mirrors. Each mirror is tilted a bit differently from the others, so that the light will strike the top of the tower.

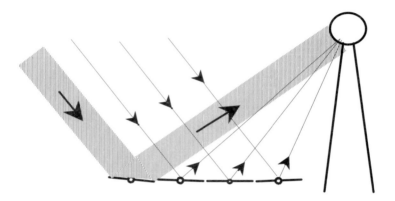

Figure 43: Schematic of sunlight reflecting from an array of mirrors to heat a central tower. Looking south at about 9:00 a.m. in midsummer.

Figure 44 shows the same mirror assembly at 3:00 p.m. The sunlight comes from the southwest and reflects toward the tower; however, some parts of the mirrors are in the shade. The drawings are exaggerations that make the situation look a lot *better* than it actually is. The distance to the farthest mirror is only a little greater than the height of the tower.

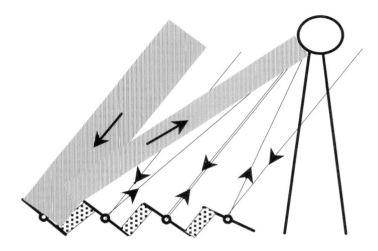

Figure 44: The same array of mirrors as shown in Fig. 43, but at 3:00 p.m. instead. The shaded area shows that some parts of the mirrors will be in shade, and some reflected light will be blocked by the backside of the nearby mirror.

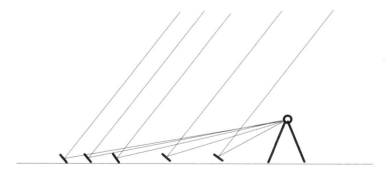

Figure 45: The geometry at Solar Two. The distance from the tower to the farthest mirrors is about four times the height of the tower, not equal to the height of the tower as shown in Figs. 43 and 44.

In the actual Solar Two installation, the distance to the farthest mirror is about four times the height of the tower, as shown in Fig. 45. Obviously, the mirrors need to be spread out considerably so as to avoid the shading problem. In the aerial photograph (Fig. 41), which shows only a small part of the 130-acre (53-hectare) installation, you can still see that the mirrors become spread out more as the distance from the tower increases.

How About a Bigger Field of Mirrors?

Solar Two is, of course, a demonstration project that is not intended to provide serious amounts of power. But how could it be scaled up? One way would be to double the diameter of the field of mirrors and to double the height of the tower. The mirrors would quadruple the power delivered to the top of the tower. Of course, the tower would have to be twice as tall and would have to carry the additional weight of the larger power-handling equipment. The legs would have to be *more* than four times as strong, not only to carry the additional load, but to provide more rigidity against wind thrust. All in all, then, there is little point in increasing the size. It would probably be just as easy to simply build another entire system, tower, mirrors, and all.

The Solar Two site occupies 52.6 hectares (130 acres) and produces 10 MWe *peak*. Its capacity factor is about 16%. For a Solar-Two installation to produce as much energy as a typical 1000-MWe power plant does in a year, it would have to cover about 33,000 hectares (127 square miles). That is environmental impact!

SEGS (LUZ International)

> "The largest solar electric generating plant in the world is the 355-megawatt LUZ International 'solar-thermal' plant, located between Los Angeles and Las Vegas, which delivers its power to Southern California Edison. Not a photovoltaic plant, LUZ is a 100-acre field of parabolic trough collectors in the Mojave Desert; these mirrors …"
>
> Berman & O'Connor (1996)

Figure 46: The solar-trough array at Kramer Junction, California. The shiny surfaces focus sunlight onto the black tubes that run along the array. The therminol fluid, thus heated, is pumped to a heat exchanger where it is used to boil water for a (Rankine) steam engine. As the sun moves from east to west, the reflectors rotate so as to keep facing the sun. Photograph courtesy of http://www.nrel.gov/data/pix/search-pix.html, taken by Warren Gretz.

Not far away from Solar Two, in Kramer Junction, California, there is another solar installation called *SEGS* (Solar Electric Generating System, built by LUZ International), of which there are nine units.[59] SEGS uses an array of parabolic mirrors (see Fig. 47), laid out on north-south axes to concentrate reflected sunlight onto a black tube through which therminol flows. The therminol delivers the heat to a Rankine steam engine whose shaft turns a generator. The SEGS installation has a natural gas boiler and a

---

[59] Gilbert E. Cohen, David W. Kearney, and Bob G. Cable, "Recent Improvements and Performance Experience at the Kramer Junction SEGS Plants," (www.kjcsolar.com/ASMEF961.html, downloaded 3/9/99).

gas reheater. The purpose of the gas assist is to assure that the steam turbine receives steam at 371 C regardless of the actual temperature of the therminol in the solar field.

Figure 47: The solar-trough array at Kramer Junction, California, showing a worker for size comparison. Photograph courtesy of http://www.nrel.gov/data/pix/searchpix.html, taken by Luz International.

Together, the nine SEGS units have about 2.3 million square meters of "aperture," the cross-sectional area of the mirror assembly facing upwards. At a full summer noontime intensity of about 950 watts per square meter, the system is exposed to about 2.2 billion watts of solar heat.

There are three processes at work, each with its own efficiency. The *optical efficiency* varies from 71% (units I and II) to 80% (units VIII and IX). That is, between 71% and 80% of the sunlight that strikes the mirrors is actually reflected to the pipes containing the therminol. They achieve this high efficiency by washing the mirrors every five or so days, and with a high-pressure wash every ten-to-twenty days. Let's repeat that: they wash several million square meters of mirror — much more than the 2.3 million m^2 of aperture — about 25 times a year!

Solar applications are low technology.

John Keyes, *The Solar Conspiracy* (1975)

The *thermal efficiency* — that is, the efficiency by which sunlight directed at the oil-containing tubes becomes heat within the system — varies from 35% to about 50%.

The *thermodynamic efficiency* — the efficiency of converting heat to work (thence to produce electricity) — is unspecified in any available literature, but is probably about 35%. One should expect an overall efficiency, therefore, of 71% × 35% × 30% = 8.7% at the minimum to 80% × 50% × 35% = 14% for the best units. That is, about 8.7% of the 2.2 billion watts of sunlight would be 190 MWe (megawatts, electrical); 14% would yield 308 MWe.

In any case, the nine SEGS units together produce about 900 GWh of electricity per year, equivalent to an around-the-clock average output of 103 MWe *with*, we add, the help of the natural-gas backup system to produce power in cloudy weather. Any serious power plant these days delivers about 10 times as much, namely, 1000 megawatts.

To qualify for the tax credits for solar installations, SEGS can use natural gas to supply up to 25% of the energy of the plant, and that's about how much they use. That is, sunlight in the SEGS system produces 75% of the 900 GWh/year produced by the system. This is equivalent to 77 MWe average power, far less than its rated power of 355 MW. The capacity factor of the plant is therefore 77/355 = 22%.

According to the National Renewable Energy Lab (NREL), the productivity of the SEGS system is about 0.5 MWe/hectare (equal to 50 W/m^2),[60] referring to *peak* MWe per unit of land area. With the capacity factor of 22%, this amounts to an around-the-clock average of about 10.8 watts per square meter. To scale that up to the size of a 1000-MWe plant would require 92 square kilometers, or about 33 square miles.

Importantly, SEGS produces its power in midday when it is needed most. As a supplemental source of electricity, units like SEGS can undoubtedly play a role *if* they can get the cost down. Solar energy is free, it is said, but the cost of electricity from SEGS is now about 8–10 cents/kWh. That's the *wholesale* price of course, and the price is that low only because of numerous tax incentives and subsidies.

If the good news is that the SEGS system (using a natural-gas backup system to maintain power) produces a whole 10% as much electricity as one

[60] http://www.nrel.gov/documents/profiles.html, "Profiles In Renewable Energy: Case Studies of Successful Utility Sector Projects"

large nuke, the bad news is that the SEGS system's paltry output is *90% of the world's direct solar electricity.*

> "But LUZ went bankrupt, a victim of the cuts in tax incentives for renewables…"
>
> Berman & O'Connor (1996)

> California's aggressive implementation of the 1978 Public Utilities Regulatory Act (PURPA), combined with 25 percent federal and 25 percent state tax credits (yielding a net after-tax benefit of 38.5 percent) allowed the company to secure a market for its first plants even at their high initial cost. … the company *tragically* went bankrupt in 1991 … [emphasis added]

> … despite LUZ's *surprising* success …[emphasis added]
>
> Michael Brower (1992)

Fire!

On February 27th, 1999, a tank containing some 900,000 gallons of the therminol caught fire at the SEGS-II unit. The fire escaped the attention of the pro-solar *New York Times* and the *Wall Street Journal*, but was reported in the *Los Angeles Times*. Until SEGS-II is repaired, the power output of the SEGS system will be about 8% less.

Photovoltaics

Of course, when the topic of solar electricity comes up, people naturally think of direct electronic conversion of sunlight to electricity without involving any heat, using solar cells. The devices for converting light to electricity are called *photovoltaic* (PV) cells. Al Gore, the best scientist ever to run against George W. Bush in a Presidential election, refers to them as "… photovoltaic cells, *small* flat panels of silicon …" [emphasis added].

It all seems to be magic. Sunlight in. Electricity out. But there's a limit to the magic. Solar cells cannot create energy. They can only convert some of the solar energy — and certainly not all of it — to electricity. Nor can the solar cell in New York convert the sunlight hitting Bangladesh into electricity. The solar cell can convert only the sunlight that the solar cell absorbs.

Yes, most solar cells are small, as Gore says. Correspondingly, they can only produce small amounts of electricity. Knowing absolutely nothing about solar-cell manufacture, nothing about their characteristics, and

nothing about their cost, we can establish an upper limit to their performance. The 1-cm by 4-cm solar cell on a calculator could produce a whopping 0.08 watts in full sunlight *if* its efficiency were (per impossible) 100%; in reality, its efficiency is between 5% and 10%, so it produces between 0.004 and 0.008 watts in full sunlight. In a brightly lit room they will produce 100 times less power. Fortunately the calculator is able to run on flea power.

> "According to Reuters, Eastwood's Tehama Golf and Country Club has 242 photovoltaic panels powering everything from the clubhouse to the golf carts, producing 32 kilowatt hours a day. Sources say Eastwood has been sending thousands of surplus kilowatt-hours to Pacific Gas & Electric Co. each year, but has never received anything in return."
>
> March Hollywood News Archives
> (Mar 26, 2001)

Clint Eastwood's generosity in sending Pacific Gas and Electric "thousands of surplus kilowatt-hours each year," worth thousands of dimes, is commendable. Of course, its effect on the tens of billions of kWh produced by PG&E every year is not exactly overwhelming. Fractionally, it's the equivalent of adding one extra person to New York City's population.

Limitations on the efficiency of photocells

People like to refer to "a quantum leap" as if it were something big. That isn't even close to the meaning of the term. Let me explain.

In an atom, electrons are confined to various states. We might visualize those states as small branches on a tree, and the electrons as birds. One branch might hold two birds and another might hold eight. Whenever a branch is full, no more birds are allowed. We will assume that the lower branches are more desirable. We will also assume that there is an overall limit on the number of birds on the tree, even though many upper branches are unoccupied. (Remember, we're not really talking about the laws of bird behavior, but making an analogy for atoms.)

Suppose that a bird from a lower branch flies away. A bird from one of the upper branches may fly down to occupy the vacancy. The bird has made a *quantum leap* of, say, two meters downward. Alternatively, given incentive, a bird might fly upward to an unoccupied branch one meter above. That, too, is a quantum leap. The important idea is that the amount of leap is a fixed quantity (two meters, one meter, …). The bird cannot occupy a branch that is 1.7525 meters down from its perch, simply because there is no branch there.

At the atomic scale, there are *energy states* that electrons may occupy, but electrons may not occupy non-states. An electron may *transit* from one state to another. In doing so, it must either absorb or release energy of the exact amount to account for the difference in energy between the two states. The quantum leap refers to *energy*, not to location.

The amounts of energy we are talking about are tiny — on the order of 0.000 000 000 000 000 000 2 joules, expressed as 0.2 attojoules (aJ).[61]

If an electron moves from one energy state of 1 aJ to another of 1.2 aJ, it requires 0.2 aJ of energy to do so. If it goes from the 1.2-aJ state to the 1.0-aJ state, it will release 0.2 aJ, possibly as heat, possibly as infrared light. The *quantum* aspect refers to the well-defined states and the well-defined energy differences. To repeat, a *quantum leap* has nothing to do with the adjective *big*.

Light, including sunlight, seems to consist of little bundles of energy called *quanta* (Einstein's terminology) or *photons* (modern terminology). Each photon has a certain amount of energy. The shorter the wavelength, the higher the energy. On the blue end of the spectrum, photons have about 3 eV, and on the red end, about 1.8 eV.

The Quantum Limitation

Photocells consist of a pair of dissimilar semiconductors joined together. The details are of no great concern to us; the only thing that matters is that photons of light striking the surface cause electrons to undergo a quantum leap. The energy gap of that leap is called the *band-gap*. That leap is the source of the electrical current that the PV cell produces.

In silicon solar cells, the band-gap is 1.1 eV. This energy corresponds to a wavelength of about 1130 nm, which lies in the infrared portion of the spectrum. Refer to Fig. 48. Photons of less than 1.1 eV energy (wavelengths longer that 1130 nm) cannot cause the transition to occur. (Think of the birds simply not having the energy to fly to the upper branch.)

If, on the other hand, a photon of 2.5 eV (about 500 nm in Fig. 48) strikes the PV cell, it has plenty of energy to cause the quantum transition. In fact, it has 1.4 eV more than necessary. The problem is that the excess energy is simply wasted as heat.

---

[61] For physicists, it would be more convenient to use the electron-volt (eV) here, but we will stick to SI units.

In Fig. 48, we show the solar spectrum (leaving out many bumps and wiggles) as the heavy curve. The 1.1 eV energy is marked on the wavelength scale as the border between having too little energy and too much energy.

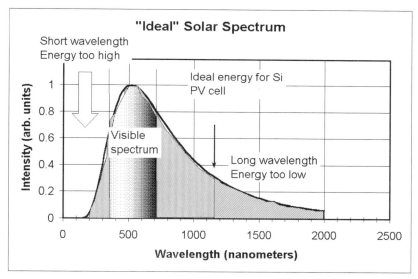

Figure 48: The "ideal" solar spectrum (no account taken of atmospheric absorption) as it applies to silicon PV cells.

Now we come to the main limitation of PV cells. We ask a broad spectrum — sunlight, ranging from infrared to ultraviolet — to do a quantum job, causing electrons to transit between well-defined states. No matter whether the transition involves 1.1 eV or 2.8 eV or any other number of your choosing, some light will always be useless because its energy is too low. Other light will always have too much energy, and the excess will be wasted as heat. For that reason, PV cells are inherently limited as to their efficiency. As a matter of practical concern, the best solar cells available on a large scale have an efficiency of about 10%.

Exotic materials!

There is a seductive fallacy about solar cells. "Silicon is one of the most abundant elements in the earth's crust. Solar cells are made from silicon. Therefore, silicon PV cells are made from readily available materials."

> "Soon to be competitive in price with other forms of energy, it [PV] is a simple, non-polluting source of energy made from readily available materials, fueled at no cost from the sun ..."

Oppenheimer & Boyle (1990)

Silicon (Si) is extremely abundant, 30% of the mass of every grain of sand in the world. That sand is not a PV cell, nor is the silicon that is extracted from the sand. But the availability of sand is not the limiting factor. To make a PV cell means to "dope" the silicon with other elements to cause the device to have semiconductor properties. When all is said and done, the Si-based PV cell has an efficiency of 10% or less. The efficiency is low because the band-gap of silicon — 1.1 eV — is so low that most of the spectral energy is wasted as heat.

To make PV cells of higher efficiency requires exotic materials, including germanium, gallium, antimony, indium, and cadmium, possibly with gold or platinum for conductors.

A study by the American Physical Society (APS)[62] asked how much material would be required to produce 1% of the US electricity by the turn of the century. They found, for example, that 250 metric tons of germanium would be required, which is over three times the world's annual production. It would require nearly twenty times the world's annual production of gallium. Some 17 percent of the US annual production of cement would be devoted to making the *structure* to hold the flat-plate solar collectors. Nothing important has changed since the 1979 APS report, save that we use 60% more electricity.

The Price Will Fall Through the Floor!

Another seductive fallacy is that mass production will cause the price of PV cells to drop the same way that the price of computing power has dropped. Solar enthusiasts incessantly tell us that the price of PV will drop through the floor, because the price of computer components has done so. Just look at how the price of computing has plummeted during the last several decades!

What does that say about photovoltaics?

Unfortunately, not a thing.

---

[62] *Solar Photovoltaic Energy Conversion: Principal Conclusions of the American Physical Society Study Group on,* H. Ehrenreich, Chairman, (American Physical Society, January, 1979).

Computers have gotten faster because manufacturers are able to put more and more transistors into smaller and smaller areas, tens of millions of them in the area covered by your fingernail. The smaller the transistors, the smaller the distance that signals have to travel, and the smaller amounts of energy that have to be stored and released. The speed of computation is inherently linked to the small size and high-density packing of transistors in integrated circuits. The price *per transistor* continues to plummet, as it has done for decades. The price *per square millimeter* of integrated circuit decreased rapidly at first, but has bottomed out.

By contrast, the name of the game in PV is *Real Estate*. To intercept huge amounts of sunlight, you have to cover huge areas with PV cells, and the size of the individual cells is of no relevance at all. The price per square centimeter of solar cells decreased rapidly at first, but has now stabilized.

It would require on the order of 15,000 square kilometers — about 20% bigger than the land area of Connecticut (but located in sunnier climes) — of PV cells at 10% efficiency — packed edge-to-edge, with no room for squirrels, let alone room for roads and houses — to collect enough sunlight to power the nation's electrical grid. At this date, the *total* amount of silicon wafer — for computer chips, TV sets, wristwatches and everything else — that has ever been manufactured in the world is in the range of a few square kilometers.

> "Photovoltaics, by contrast, are already here, in commercial quantity."
>
> Oppenheimer & Boyle (1990)

If by "commercial quantity" Oppenheimer and Boyle were referring to PV cells for calculators and battery chargers in remote telephones, they would be correct. However, they have more grandiose things in mind, like competing "compete with fossil-fuel electricity sources at a reasonable price."

In the US, the total shipments of PV cells manufactured from 1982–1998 amounts to 275 *peak* MWe. Assuming an average efficiency of 10%, this implies a total surface area of 2.9 square kilometers. The *peak* power of 275 MWe amounts to about one fourth of the power produced by a conventional power plant, and the year-round average (optimistically 55 MWe) is, shall we say, pathetically small compared to anything useful.

> "The United States alone could be producing as much as 1000 megawatts in photovoltaic capacity by 2000 at costs competitive with peak-hour electricity from conventional sources."
>
> Paul & Anne Ehrlich (1991)

In 1997, we produced 46 MWe of PV capacity, a mere 4.6% of what the Ehrlichs said we "could be producing."

Suppose that we wanted to generate enough power from PV cells to be equivalent to 1000 MWe of electricity around-the-clock — equal to one large conventional power plant — using PV cells of 10% efficiency and year-round average sunlight at 200 W/m^2. We would require 50 million square meters (20 square miles) of solar collector. This is about twenty times the total US production of PV cells in the nine-year period 1982–1998 (inclusive).

> "The *only* factor that inhibits the wider use of solar cells is the cost of manufacturing them." [emphasis added]
>
> Oppenheimer & Boyle (1990)

> "The limiting factor is *not* simply the cost of producing the photovoltaic cells." [emphasis added]
>
> Berman & O'Connor (1996)

The American Physical Society study [Ref. 62] concluded:

> "Because of the costs associated with encapsulation, foundations, support structure, and installation of PV array fields, there is a large economic penalty for use of low efficiency cells. To compete in the U.S. central power generation market, *even zero-cost PV cells must have a limiting minimum efficiency*. The use of modules with efficiency as low as 10% will probably require substantial reduction in these other costs, even if the modules themselves are inexpensive." [emphasis added]

The APS concluded, in other words, that the cost of solar PV electricity would be prohibitively expensive even if the cells themselves were *free* ("zero-cost"), unless the efficiency were high.

The APS does not even mention the costs associated with storage, because they recognized that "it is unlikely that photovoltaics will contribute more than about 1% of the US electrical energy produced near the end of the century." [Ref. 62.] Good numbers and good physics led to good predictions.

> "Two factors determine the high price of photovoltaic power: the efficiency with which a square foot of cells converts sunlight into electricity and the cost of producing that square foot."
>
> Oppenheimer & Boyle (1990)

As we have seen, they determine *part* of the high cost. Absolutely none of this was news in 1989 when H. M. Hubbard made an optimistic prediction that we encountered in Chapter 3 and repeat here:

> "Photovoltaics has the advantage of a relatively short time period (1 to 2 years) required to bring a large (1-GW[$_e$]) power plant on line ... researchers at Chronar Corporation have optimistically estimated ... approximately 10 GW[$_e$] per year of PV manufacturing capability could be in place by 1995, leading to *40-GW[$_e$] installed capacity by 2000*." [emphasis added.]
>
> H. M. Hubbard (Former head of SERI) (1989)

Use the Deserts!

The best place in the US for solar collectors of any sort is the desert Southwest, if we ignore the difficulties in transmitting power to other parts of the nation. That is, of course, why both Solar-Two and the SEGS units are in the Mojave Desert. Brown, Flavin, & Postel discuss putting huge solar collectors in the desert, and shortly thereafter say,

> "Decentralization may be another hallmark of the emerging new energy system."
>
> Brown, Flavin & Postel (1991)

With respect to "decentralization", there is little to distinguish large companies producing energy in the desert environment from large companies producing energy by some other means elsewhere.

If there were large energy-producing facilities in the deserts, how would the energy be delivered to New York, Boston, Washington, D.C., Philadelphia, Atlanta, and Chicago, thousands of kilometers distant? High-voltage power lines? Through whose neighborhood?

Put 'em on the Roof!

> "Solar technologies need not be spread over large swaths of land. Photovoltaics can be deployed on rooftops.
>
> Brown, Flavin & Postel (1991)

There are about 100 million homes in the United States. For the sake of argument, let us assume that every one of them occupies 150 square meters of land area. That amounts to 15 billion square meters. Hypothetically, if *every* roof were fully covered with solar cells at 10% efficiency, every square meter could generate a year-round average of 20 watts per square meter. That amounts to 300 GW_e.

Life isn't that simple. Apartment dwellers have virtually no roof area available to themselves. Houses usually have a roofline that extends north-south or east-west. The houses best suited to solar collectors are those with a southern exposure.

But there are more problems. Most houses are in cities or suburbs. A characteristic of people in such places it to plant trees, often deliberately on the south side of the house to shade the house in summer. In the largely treeless west, you can spot towns miles away simply because they have trees.

All things considered, rooftops on American homes might be able to generate a year-round average of about 10 to 20 GW_e. Even that possibility is entirely hypothetical, of course. Most people cannot afford to cover even part, let alone the entire south-facing roof with solar collectors.

Solar Cars!

Every year engineering schools have a race to determine which school's students produce the best solar cars. The students learn a lot from the exercise, especially in how to reduce the weight and air resistance of vehicles, and how to make the most of the very little power they have at their disposal. It is a good engineering project that brings out the best engineering talents in the competitors.

Every single student learns by direct experience that the world's transportation system will not be based on solar cars and trucks. They also learn that if the efficiency of the PV cells were 100%, the available power would still be too small for any practical vehicles. Invariably, the solar cars are small, cramped, uncomfortable, slow, extremely dangerous in case of accident, and devoid of amenities. They run only where and when the sunlight is intense; no Alaskan will use one in February to go out to get a bottle of milk.

Ideas for the future

Solar cells that are large enough to be useful for solar applications are about 10% efficient. Small experimental PV cells made of exotic materials have achieved much higher efficiencies, some in the range of 20%. It is to be hoped that high efficiency can be obtained for solar cells that can be mass-produced so that large areas can be covered cheaply.

The primary method of attack is to make the PV cell in layers. The layer facing the sun uses blue light, but is transparent to all other colors. The next layer down absorbs green light and is transparent to all longer wavelengths, and so on. Everything is engineered so that if a given amount of sunlight causes a million transitions in the "blue" layer, it will also cause a million transitions in each other layer. The current must be the same in all layers, and will be limited by the current in the weakest layer.

This Holy Grail of photovoltaics has remained elusive. Be reminded that the most democratic institution on earth is the periodic table, found in classrooms throughout the world. This list contains the elements — *all* of them — that exist for constructing PV cells (or anything else). The reward for developing reliable, inexpensive, high-efficiency PV cells will easily be in the billions of dollars, as everybody has recognized from the outset. Still, despite many decades of corporate and university research, such PV cells have not been developed.

Chapter 12. Miscellaneous Non-Solar "Renewable" Sources

There are some renewable sources of energy that are not actually soalr energy. Geothermal energy arises from the decay of radioactive elements deep within the earth, and has nothing to do with sunlight. There is a virtually infinite supply of geothermal energy, but it is very hard to use.

Oceanic waves are due to winds (indirectly due to solar energy), but tides are due to the relative motions of earth, moon, and sun, and not due to sunlight. The high tides at a few well-known tidal basins like the Bay of Fundy are due to the geometrical features of the water channel.

Although these sources are considerably different from those discussed in the rest of the book, we should discuss them, albeit briefly.

Geothermal Energy

The source of geothermal energy is radioactive materials found throughout the volume of the earth. The temperature of the interior is high because of that radiation and because the earth's crust provides an effective barrier against the transmission of heat.

Generally, the deeper you drill into the earth, the hotter it gets. However, there are places where hot magma is not very deep at all; Hawaiians can go watch lava ooze out of the ground. Icelanders get most of their energy — both heat and electricity — from geothermal sources.

There is enough energy in the bowels of the earth to last the human inhabitants forever. But how can we *use* that energy? Our techniques are primitive, but it is certain that our technology will improve.

When we extract energy from *anything*, the energy of that source diminishes, and geothermal hot-spots are no exception. The geothermal power plants in Middletown, California's misnamed "geysers region" produced 2000 MWe of electricity — the equivalent of two large conventional power plants — in the 1980s, but can now produce only 850 MWe.[63]

The diminution of power is not a reason not to develop geothermal power, of course. What it means in practical terms is that the hunt for more

---

[63] Jennifer Coleman, "Running out of steam: Geothermal field tapped out as alternative energy source," Associated Press (*Pueblo Chieftain*, Apr. 15, 2001).

geothermal energy is a never-ending project, not unlike the quest for petroleum.

Waves

Oceanic waves are also an indirect source of solar energy. The source of energy for the waves is the winds, blowing over hundreds or thousands of kilometers, whipping up the water. The physics is complicated, because the height of the waves can't be predicted from the frequency, and conversely. *Both* distributions are Rayleigh distributions[64] not unlike those we describe for wind in the Appendix B, "Wind-site Predictions." Additionally, the power in the waves when they reach shallow water is no more than about a quarter of the deep-water value, and the efficiency of converting low-head slow wave energy into electricity would be perhaps 25% to 30%. That is, we might expect about 25% of 25% (one part in 16) of the wave energy to be realized in the form of electricity.

The average wave power (greater than the electrical power that might be extracted) along the coasts is in the vicinity of 30 MW/kilometer.[65] One fourth of that is just under 8 MW/km.

Wave Energy in Realityland

There was a small, 24-foot long installation with floats that would rise and fall with the waves off the coast of New Jersey in the early 1900s that produced about 1.5 kW. This translates into about 1.2 MWe per kilometer of coastline, not far from our estimate of 2 MW$_e$/km. If the *entire* coastline on both Atlantic and Pacific coasts were to produce power at that rate, we would generate about 6000 MW$_e$, less than 2% of the average US electrical energy production. It is highly doubtful that environmentalists (or surfers or boat-lovers) would approve the project.

---

[64] C. L. Bretschneider, "Sea Motion," in Handbook of Ocean and Underwater Engineering, (McGraw Hill, New York).

[65] Peter Weiss, "Oceans of Electricity: New technologies convert the motion of waves into watts," *Science News* **159**, pp. 234-236 (Apr. 14, 2001) gives 40-70 MW/km for a few excellent sites and 10-20 MW/km for southeastern US.

Tides

> "… and we are seeing the re-emergence of tidal power."
>
> Al Gore   (2000)

The causes of tides are more subtle. The gravitational pulls of the moon and of the sun cause very small perturbations in the sea level,[66] and the rotation of the earth causes those perturbations to move with respect to the continents. These "tides" are very small.

There are certain tidal basins, including the Bay of Fundy and the Rance Estuary (in France), where tides get very high. In such places, several effects contribute to the high tides. One is that of resonance. Given a tendency for water to run in and out of a basin, there is always some natural frequency for doing so. Whenever the frequency matches that of the solar/lunar tides, the amplitude of the tide can build up.

The effect can be seen in any pool of water, such as a sink or bathtub. Push down slightly with the open hand and a wave moves away. When the wave reflects off a wall and comes back, lift up the hand. As the water starts to leave, push down slightly again. Before long, there is a large wave. What makes it work is timing. Whether the water is coming in or going out, you are always adding energy to the system. This can happen only when the frequency of the driver (your hand) matches the natural frequency of the system (the water in its container), that is, when there is a resonance.

Another phenomenon is the funneling of water. Suppose that westbound water has a tide that is no more than a centimeter high, but is a "wave" that is 500 kilometers in length. As it moves toward a continental coast, if the water gets funneled into a channel that is only 1 kilometer wide, all of that excess water — the one-centimeter high wave — becomes a wave that is 500 times as high, namely, 500 cm, or five meters. Moreover, the wave speed increases, so the high tide comes in rapidly.

The inverse of this funneling phenomenon is well known to Long Island Sound boaters. At the eastern end of Long Island, there is a narrow passage for the tide to pass through, known as *The Rip*. When the tide is going out, it is a real chore to be able to get a boat into the Sound. A similar phenomenon happens at San Francisco Bay.

Tidal Basins in Realityland

This discussion of tides is certainly incomplete, but our purpose here is merely to show why there are very few tidal basins of note in the entire

---

[66] More correctly, the *gradients* in the gravitational pulls are responsible.

world. High tide occurs twice a day. That is, for a short time, a tidal-hydro station would be operating at full head (far less than the hydraulic head of major hydropower dams). As the water runs out, the head decreases.

There is an existing 240-MWe (peak) power plant at the Rance Estuary in France that produces 500 million kWh per year, corresponding to a capacity factor of 26%.[67] There was a plan in 1961 to develop Passamaquoddy Bay (off the Bay of Fundy) between Maine, New Brunswick, and Nova Scotia, where 15-meter tides exist. It would have been a 230-MWe power plant, probably with a similar capacity factor of around 25%. There are, in fact, only a few such sites in the world, so the effect on worldwide energy production is obviously trivial.

Non-Solar Non-Source: Hydrogen

A few decades ago, it was common to speak of the "hydrogen economy," wherein hydrogen would be produced by electrolysis of water, with the energy being supplied by nuclear or fusion reactors. This would be a way to use energy from stationary power plants into a form that could be transported. Everybody recognized that:

(A) there are no hydrogen wells; hydrogen is not a *source* of energy;

(B) energy is *required* to produce hydrogen and that it would be impossible to get all of that energy back;

(C) storage and transportation of hydrogen are not trivial matters; and

(D) while not a hazard beyond reasonable control, hydrogen is inherently more hazardous to handle than natural gas.

Today, unfortunately, it is common to hear about a "hydrogen economy," wherein hydrogen is regarded as a *source* of energy. Cars will run on fuel cells, using hydrogen as an *energy source.*

Hydrogen is no more a source of fuel than the refrigerator is a source of milk or the electric socket is a source of electricity. But this simple fact doesn't keep people from claiming the contrary.

> "The third age of fuel: Just as coal gave way to oil, oil may now give way to hydrogen."
>
> *The Economist* (Oct. 25, 1997)

---

[67] Gordon L. Dugger, "Ocean Thermal Energy Conversion," in Douglas Considine's *Energy Technology Handbook* (McGraw-Hill, 1977).

"Hydrogen: Tomorrow's Petroleum"
Headline at Worldwatch website,
http://www.worldwatch.org/alerts/010517.html

"Hydrogen is perhaps THE ideal fuel known to man at this time. It is a fuel source that has no chance of running out until our Sun stops producing it."
"Hydrogen: The Neverending Energy Source,"
http://www.tcnj.edu/~energy/altfuel/Hydrogen.htm

In one sense this individual from an educational institute is correct. We can't run out of something we don't have.

"The U.S. Department of Energy supports world-class R&D in hydrogen technologies that will provide America with near-, mid-, and long-term strategies for a clean, sustainable, domestic energy *supply*." [emphasis added]
http://www.eren.doe.gov/hydrogen/

"Renewable Energy Technologies: Hydrogen"
Headings at http://www.eren.doe.gov/RE/

Using Solar Energy To Produce Hydrogen

Researchers have discovered how to electrolyze hydrogen from water using solar energy and PV cells that produce just the right voltage;[68] the efficiency is about 12.5%. The photovoltaic cells are made of GaInP and GaAs.

Other researchers have discovered algae that can produce hydrogen as a byproduct. They don't even mention the word *efficiency*, but we can make a reasonable estimate. One liter can produce 3 milliliters of hydrogen per hour in bright sunlight.[69] Let us assume that the liter is a cube, 10 centimeters on a side. At a solar intensity is 950 W/m^2, the rate at which solar energy enters the liter is 950 W/m^2 ×0.1 m × 0.1 m = 9.5 watts. The 3 ml has a mass of 0.26 mg; its heat content is 37 joules. The rate of

---

[68] Robert F. Service, "A Record in Converting Photons to Fuel," *Science* **280**, p. 382 (17 April, 1998).
[69] C. Wu, "Power Plants: Algae churn out hydrogen," *Science News*, p. 134 (Feb. 26, 2000).

production is therefore 37 J/hour = 0.01 watts. The efficiency is therefore about 0.1%.

Hydrogen can be used as a carrier of energy, and sunlight can be used to produce hydrogen by electrolysis or photosynthesis. The efficiency, however, is nothing to get excited about.

Chapter 13. The Solar Fraud

Scientists and engineers have long understood sunlight and its uses. There is nothing new about the windmill formula, about the hydropower formula, about the production rates for biomass, or about the existence of PV cells. The intensity of sunlight — even that reaching the upper atmosphere — has been understood for well over a century. Efficiency has been a hot topic for two centuries. Using mirrors to concentrate sunlight has been known for millennia.

There is an old adage that if something sounds too good to be true, it probably is. That adage is especially applicable to solar energy. For decades, there have been delirious proclamations that world would soon run on solar energy. Those statements have always sounded too good to be true and, sure enough, have always been false.

Every year Lucy promises to hold the football so that Charlie Brown can do a place-kick. Every year as Charlie charges toward the ball, Lucy pulls the ball away at the last minute. Every year Charlie Brown lands on his back. Every year solar sirens like Carol Browner and Denis Hayes tell us that solar energy is the answer to our problems. Every year suckers fail to learn from experience.

Hope springs eternal. The news media continue to publish and broadcast — unquestioningly — glowing stories about solar homes, wind farms, and cow manure, despite decades of failed predictions made by solar enthusiasts. Apparently, the enthusiastic statements *do not* sound too good to be true to journalists and politicians. Such is the price to be paid for Pablum-level education in science.

Low-Tech, High-Tech

Broadly speaking, there are two ways to use solar energy.

In the traditional ways, the solar energy is collected by natural processes on earth. Sunlight evaporates water; we use hydropower. Sunlight produces chlorophyll; we burn trees and paper (from trees), and methane from human and animal waste. Sunlight causes the air to move; we put wind turbines where the winds are strong.

But we can also collect sunlight with man-made devices. Greenhouse-type solar heat collectors warm some homes and swimming pools. Focused-light devices (*e.g.,* Solar-Two and SEGS) can produce high-temperature heat to run steam turbines. Photovoltaic collectors produce electricity from sunlight.

Natural Collection & Storage of Energy

As it happens, we get most — indeed, almost *all* — of our solar energy from two of those natural sources. Biomass and hydropower together provide only about 7% of our energy. The average yield of firewood in New England's abundant forests is about 0.12 W/m^2, and that 1.2 W/m^2 is considered very high yield. The net yield from ethanol produced from highly fertilized corn yields a trifling 0.05 W/m^2 under the very best of conditions, but is usually negative instead. Despite the low energy intensity of biomass, it provides us with about 3.2% of our energy.

Wind ranks with biomass in terms of energy intensity, producing about 1.2 W/m^2 of land area in wind farms. Nevertheless, wind produces a minuscule one one-thousandth of our electricity. Wind power is currently the fair-haired child of the renewable-energy line-up, but has come to its stature through massive subsidies. Utilities offer their customers the opportunity to purchase "green" wind power for an *extra* several cents per kilowatt-hour, but generally would prefer to shun the low-quality power if given the chance to do so. For better or for worse, wind is the best solar energy available, aside from biomass and hydropower.

Collection & Storage of Energy by Manmade Devices

The solar energy collected by man-made devices involves higher energy intensity. Solar-2 produces year-round electrical energy at the average rate of about 3 W/m^2. LUZ International's SEGS project produces electrical energy at the average rate of about 11 W/m^2. Today's commercially available PV cells produce power at a year-round average of perhaps 20 W/m^2 if the load is continuously adjusted to keep the PV cells operating at their peak efficiency. Solar heat collectors for homes deliver heat at the year-round average rate of about 60 W/m^2.

Clearly, man-made solar collectors are greatly superior to natural ones, as indeed they ought to be; man has his own interests at heart. But they can collect only the solar energy that strikes them. To collect huge amounts of solar energy means building huge amounts — tens of thousands of square kilometers — of solar collectors. As a result, all such high-tech collectors, combined with all wind turbines, produce about one part in 850 of the energy we use.

The high-tech direct-sunlight producers of electricity are expensive projects, and high-maintenance ones to boot. Mirrors and PV cells need frequent cleaning. The to-date total world production of PV cells is still in the range of a few square kilometers, even two decades after the computer revolution began. The Union of Concerned Scientists (1991), by implication from Table 1 had expected about 1000 square km of PV cells to be in operation by the year 2000.

Non-Sources of Energy

Conservation and hydrogen have three things in common. Neither one is solar energy. Neither one is a source of energy. Both have adherents who claim that they *are* sources of energy.

Perhaps they are just playing with words. Still, if the proponents are dumb enough to say such things, it follows that there are people dumb enough to believe them.

Delusional Economics

[handwritten: what about oil subsidies ???]

The Soviet Union had an army of bureaucrats to determine — by fiat — the prices of millions of items ranging from machine screws to bread, from picture frames to hobnails. The United States has millions of people to determine the prices of goods and services. They vote with their pocketbooks. The US has a robust economy (even in crises) and the Soviet Union had a corrupt system that turned into corrupt chaos after the komissars were driven from power.

Governmental intervention in the market place comes with the arrogance that the bureaucracy has a better way to set prices than does the marketplace. Let us look at just a few examples of governmental meddling, aided and abetted by acolytes in the press.

Andrew Bridges[70] tells us that in "sun-kissed California, the energy source that once languished on the economic fringe is now carving out a booming niche among consumers ..." P. T. Barnum was right. The story comes with a picture of a California University professor sunning herself beside her beloved photovoltaic cells on her roof.

The caption to the picture reads, "Buoyed by generous government subsidies and plummeting costs, solar energy is enjoying a rare day in the sun." The first part of that claim is certainly correct.

The good professor says, "My meter runs backward during the day." Translation: She sells electricity to the utility at retail rates. What a neat way to capitalize on the government money that was responsible for the photocells in the first place. Oh, and when the milkman comes to deliver a quart of milk, we'll make him buy a quart of our milk instead — the quart we got with our food stamps.

---

[70] Andrew Bridges, "California turns to sun to ease power woes," Associated Press, 8/7/01.

I quote from the web site of the *Business Journal of Portland.* "The Oregon Public Utility Commission approved a proposal for the utility to collect a $5 fee from customers who opt to purchase one or more power 'blocks,' [of wind energy] which represents 100 kilowatt-hours of electricity." That amounts to a direct payment of *an extra* $0.05 — *an extra* nickel — per kilowatt-hour. No, that's not a subsidy, because people are given an option. However, it speaks volumes about the cost of wind energy.

Oregon has approved a "net-metering" law that amounts to a subsidy to people who install solar toys and an outright punishment to the utilities. To quote the Renewable Northwest Project (RNP), "'Net metering' allows customers who own small renewable energy systems or fuel cells to offset electricity purchases from their utility by running their meter backwards while their small-scale sources are at work. At the end of each billing period, consumers pay for only the 'net' energy purchased from the utility."

The fuel cost of producing electricity is in the range of 2 cents per kWh; that we pay about a dime per kWh has to do with the entirety of the utility's expenses, everything from amortizing equipment to maintaining power lines. When gung-ho pro-solar parasites sell electricity to the utility for a dime per kWh — as they do in "net-metering" — the utility is forced to subsidize rich people's toys to the extent of eight cents per kWh.

Boulder, Colorado, like many college towns, is a rest home for teenagers and a haven for self-righteous Protectors of the Earth. An article[71] in *The Denver Post* on June 17, 2001· brings us the good news about solar energy in a politically correct subdivision not terribly far from the National Renewable Energy Laboratory (known locally as La-La Land). You'll be happy to know that prices for single-family homes "start at under $300,000 — but for another $7500, the builder will add a *small* [emphasis added] photovoltaic power system that at max performance can deliver 550 watts — enough to power 10 [55-watt] lightbulbs." In full sunlight, of course, just when you need it. Samuelson (footnote 71) refers to that as a "low-cost upgrade."

Gee, I was more interested in running a 5000-watt clothes dryer. I think I'd better get 9 of these 550-W systems. It will only cost $67,500. Why, that's even less than the cost of the house!

Of course, the article mentions that "the system lets you spin your electric meter backward." Welcome to Parasite Heaven.

---

[71] Mark Samuelson, "Make your own electricity? McStain lets you do it now," *The Denver Post*, (June 17, 2001).

Meanwhile, the Colorado Public Utilities Commission has ordered Xcel Energy to buy power from a wind farm to be built (with corporate-welfare subsidies to Enron) near Lamar. If Xcel loses money on the deal, they can go back to the PUC and request a rate hike so that the government-induced losses can be recovered from ratepayers. PUC gets credit for being Green, and Xcel will get the blame for higher costs.

Land Usage

Solar energy is inherently dilute, and inefficiencies in collection and distribution reduce the energy intensity even further. The more dilute the collection of solar energy, the larger the land area that is required for any given task.

To emphasize that point, let us look at the amount of land area that would be required to produce all of the 101 EJ of energy the US uses every year. The land area required would depend upon the efficiency of collection or, equivalently, upon the collected energy intensity in watts per square meter as shown in Fig. 49. If, for example, we could collect solar energy only at the rate of 0.4 W/m^2, it would require 100% of the US land area to collect all of our energy.

On the other hand, if we could collect solar energy at 20 W/m^2 — twice the output of the field of parabolic reflectors in the Mojave Desert — we could get our energy from 2 percent of our land area; that value is just under the land area of Minnesota. (Minnesota is chosen for its size, not for its sunlight.)

Even if it were possible to collect solar energy at 100% efficiency — and I emphasize that it is *utterly impossible* — the required land area would exceed that of Connecticut by some twenty percent.

On a map, Minnesota and Connecticut can be made to look small, even tiny if it suits the purpose. But the matter under discussion is the allocation of land area to solar projects. Imagine laying 4,400,000,000 — 4.4 *billion* — sheets of plywood, which is pretty cheap stuff, edge-to-edge, leaving no room for roads, houses, businesses, rivers, trees, or lakes. That's enough plywood to cover Connecticut.

Increase that by 20% to some 5.2 billion sheets of plywood, and you have enough collection area to collect all of the energy we use in the US — *if* the collection efficiency were 100%, and *if* our energy demand somehow didn't grow during the time all of those collectors were being laid out. The annual production of plywood is somewhere in the range of 0.5 billion sheets, only about 10% of our requirements.

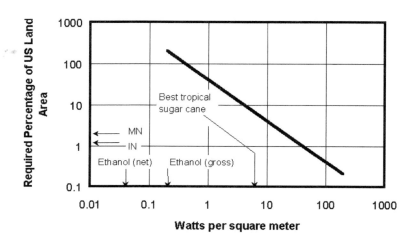

Figure 49: The required land area (heavy line) to produce all US energy (about 100 MJ/year) versus the energy intensity. The land areas of Minnesota (MN, 2.8% of US) and Indiana (IN, 1.2% of US)) are shown by the arrows on the left. The gross production of ethanol is about 0.2 W/m^2, and the net production (using the world's best equipment) is 0.047 W/m^2.

But plywood is not a solar collector of any kind, let alone one with 100% efficiency. We have simply ignored the problems of fastening the hypothetical solar collectors so that they don't blow away in the wind. We have made no provisions for cleaning the collectors of dirt or snow. We have made no provisions for delivering the energy, or for converting it to a useful form. We have merely dreamed our way through the problems of energy storage.

Oh, and who is going to file the Environmental Impact Statement?

Environmental Opposition

In this dream world of 100% efficiency and plywood substituted for solar collectors, we still are faced with an enormous project and an environmental impact of biblical proportions. Highway projects, for example, are often held up for decades because the highway might destroy wetlands that amount to only a few hundred or a few thousand sheets of plywood. Our project requires the land area of *billions* of sheets of plywood.

The easiest ways to increase solar energy would be to burn more trees and (with more difficulty) to put hydropower dams on all streams. Both attempts are met with vigorous political opposition by *environmentalists* — and not without some justification.

PRESERVE OUR
FORESTS

That is only the beginning. Environmentalists regularly oppose nuclear power plants and coal-fired power plants. They battle against natural-gas power plants and hydropower plants. They show up to protest geothermal power plants, wood harvesting, and high-intensity farming (with fertilizers and pesticides). To them, gas wells, oil wells, coal mining, gas pipelines, oil pipelines, electrical transmission lines are insults to the environment.

For one example, a utility in Hawaii[72] is attempting to enlarge the capacity of some existing high-voltage transmission lines. Exactly six of the towers are in a game preserve, as they have been for years. The utility has already spent millions of dollars on an environmental impact statement pertaining to hanging a few more wires on those six towers.

That is, environmentalists oppose every means of producing energy and every means of transmitting energy. Why would they feel good about huge energy farms? Why would they support transmission lines to carry electricity from sunny Arizona to New York? Why would they approve the use of hundreds or thousands of square kilometers of land to produce electricity? Why would they favor tens of thousands of miles of pipelines to carry hydrogen all over the country?

Make no mistake about it. The pro-solar crowd will eventually discover that solar energy projects that are big enough to be useful are big enough to be harmful to the environment. Meanwhile, they intend to kill all of the alternatives, all of the conventional ways of producing energy, all of the conventional ways of transporting energy, and all of the traditional ways of using energy.

---

[72] Alan Lloyd, Professional Engineer, long-time resident of Hawaii, private communication

The Echo Chamber

The widely disseminated anti-nuclear misinformation described in "Statistical Smoke Generators" is a metaphor for the echo chamber into which solar energy pundits yell. One person makes a proclamation and then somebody repeats it. Eventually the somewhat modified comment comes back to the person who originated it; he now believes it must be true because everybody is saying it, though in different ways. In this self-feeding echo chamber, fact yields to fiction, reality to delusion.

Journalists, while not the only problem, exacerbate the problem by acting as the powerful amplifiers of the echo chamber. They don't distinguish between conservation and sources of energy, between insulation and solar collectors, or between steady power of conventional power plants and stochastic power from wind turbines. They have no idea how many ways we use energy, nor where we use it. They have no idea that a car driving at highway speeds is consuming power at the rate of ten-to-twenty clothes dryers.

Schools of journalism teach their students to ask the traditional questions. Who? What? When? Where? Why? How?

Often the most important question is *How Much*? Schools of journalism do not teach their students to ask this question, let alone *how* to ask it. It is all too easy to read from the teleprompter that a wind farm will produce enough power for 7,500 homes. How much power *in watts* will that project produce on the average? How many *watts per square meter of land area*? What is the expected capacity factor? How does the power compare with that of a nearby conventional power plant? How many wind turbines would be required to produce as much power as that conventional plant?

Journalists serve as the mindless amplifiers in the echo chamber, but bogus signals are the root cause. We've heard from some of those folks.

"The idea that solar energy is only a pious hope for the
future, pie in the sky, is a myth — a dangerous deception."
Barry Commoner (1978)

Appendix A

The international system of units (*Systéme Internationale*, SI) is used by scientists and engineers throughout the world, including the more civilized parts of the United States. The reasons for using SI are best elucidated in Chapter 5. It is of interest that all of the units in the British system are defined in terms of SI units, not the other way around.

This appendix presents straightforward information in SI units, and provides conversion factors for readers to convert *to* SI units. Readers who insist on converting *from* the SI system to their preferred parochial units will have to do that by themselves.

The metric prefixes are given in Table A1.

Table A1: Metric Prefixes

E (exa)	10^{18}
P (peta)	10^{15}
T (tera)	10^{12}
G (giga)	10^{9}
M (mega)	10^{6}
k (kilo)	10^{3}
m (milli)	10^{-3}
μ (micro)	10^{-6}
n (nano)	10^{-9}
p (pico)	10^{-12}
f (femto)	10^{-15}
a (atto)	10^{-18}

Many units have been defined for energy. Table A2 presents conversion factors. Note that the British Thermal Unit (BTU) is actually defined in terms of the joule by a 12-digit conversion factor.

Table A2 : Energy Conversion Factors

Multiply Number of ▼	By ▼ to get joules
Watt-seconds	1
British Thermal Units (BTU)	1055.05585262
Quadrillion BTUs (quads)	1.055×10^{18}
Kilowatt-hours (kWh)	3.6×10^6
Horsepower-hours	2.69×10^6
kilocalories (kcal)	4186.8
calories (cal)	4.187
Therm ($=10^5$ BTU)	1.054×10^8
Foot-pound	1.36
Erg	1×10^{-7}
Watt-year	3.16×10^7

Often used incorrectly as energy units are the heat contents of fuels. For example, we endlessly hear of nuclear weapon yields expressed in tons of TNT. Or, people will say that such-and-such project will displace a million tons of coal. Table A3 gives heat contents of fuels by mass, and Table A4 gives the heat content by volume. One kilogram of petroleum has a heat content of about 45 MJ/kg, and the same number holds for all petroleum products. However, they have different densities. A gallon of propane weighs less than a gallon of gasoline, and therefore it has less heat content.

Table A3: Heat Contents of Fuels (By Mass)

Multiply Number of ▼	By ▼ to get joules
Kg of petroleum	45×10^6
Kg of coal	$(15 \text{ to } 30) \times 10^6$
Kg of coal (US average)	24.1×10^6
Kg of drymatter (biomass)	15×10^6
Kg of hydrogen	140×10^6
Kg of methane (CH_4)	19.9×10^6
Kg of ethanol (EtOH)	30×10^6
pound of coal (average)	10.9×10^6

Table A3 (continued)	
Multiply Number of ▼	**By ▼ to get joules**
ton of coal (average)	21.9×10^9
Kg of TNT	2.1×10^6

Table A4: Heat Contents of Fuels (by volume)

Multiply Number of ▼	**By ▼ to get joules**
Barrels of crude oil	6.12×10^9
Barrels of aviation gasoline	5.326×10^9
Barrels of motor gasoline	5.542×10^9
Barrels of propane	4.047×10^9
Barrels of kerosene	5.982×10^9
Gallon (US) of gasoline	131×10^6
Gallon of ethanol (EtOH)	95×10^6
Cubic feet of natural gas	1.092×10^6
Ton of TNT	4.2×10^9
cords of wood (white oak)	$31. \times 10^{10}$

The SI unit of time is the second; however, people frequently use other units of time. Everybody understands the length of a day or an hour; however, for calculations, one should use seconds. Table A5 shows easy conversion factors. (How many times would you like to multiply $60 \times 60 \times 24 \times 365.25$ to find the number of seconds in a year?)

Table A5: Time Units

Multiply Number of ▼	By ▼ to get seconds
Minute	60
Hour	3600
Day	86,400
Year	$3.16 \square 10^7$ ($\square$ $\square$ $\square$ 10^7)

The SI unit of length is the meter (*not* the centimeter, *not* the millimeter, *not* the kilometer). Americans use inches, feet, yards, and miles. Table A6 gives the conversion factors to convert lengths to meters. Note that the inch is *defined* to be 0.0254000... meters.

Table A6: Length Units

Multiply Number of ▼	By ▼ to get meters
centimeters	0.01
inches (definition)	.0254000000000000
feet	0.3048
yards	0.9144
kilometers	1000
miles	1.609

All areas, regardless of the shape of the surface, are ultimately found by multiplying length by length (and sometimes adding results of partial areas). Therefore, all areas in SI units are in square meters (m^2). Table A7 can be used to convert areas in common usage into m^2.

Table A7: Area Units

Multiply Number of ▼	By ▼ to get square meters
cm^2	$1 \square 10^{-4}$
$inches^2$	$6.452 \square 10^{-4}$
ft^2	0.0929
hectare (ha)	$1.0 \square 10^4$
acre	4047
mi^2	$2.59 \square 10^6$

Similarly, all volumes are ultimately found by multiplying length by length by length. Therefore, all volumes can be expressed in cubic meters (m^3); indeed, that is the only SI unit of volume. Table A8 shows the conversion factors to convert to m^3.

Table A8: Volume units

Multiply Number of ▼	By ▼ to get cubic meters
in^3	$1.639 \square 10^{-5}$
gallon (= 231 in^3, by definition)	$3.785 \square 10^{-3}$
liters	0.001
barrels	0.159

We have but one sun, and its properties are known. Solar intensity values are given in Table A9. The figure of 1367 W/m^2 for sunlight reaching the earth's outer atmosphere was known within 10% over a century ago, long before satellites measured the intensity directly.

Table A9: Solar Intensity Values

Solar Flux	Watts/m^2 (of land area)
At earth's orbit	1367
At surface, noon, tropics, clear skies	950
Maximum conceivable 24-hour average, at equator, no clouds, at equinoxes	300
Albuquerque, New Mexico, yearly average	240
US, around 48 states, around-the-year, around-the-clock average	200
Hartford, Connecticut yearly average	160

Conversion of sunlight into useable energy is not always as efficient as some people would like to believe. Table A10 gives real and theoretical production intensities (in W/m^2) for biomass, wind, and hydro.

Table A10: Solar Production Values

Solar Production	Watts/m^2 (of land area)
New England forests	0.12
US farm crops (edible portion)	0.11
Corn (whole plant)	0.75
Corn (edible)	0.25
EtOH from corn (gross)	0.195
EtOH from corn (net, best conditions achieved)	0.047
Biomass (in noon sun)	25
Sugar cane (whole plant, tropical conditions, plenty of fertilizer and pesticides)	3.7
Biomass (oceans, phyloplankton)	0.62 to .074
Biomass (maximum theoretical, average yearly production)	13.2
Hoover Dam (average power divided by collection area)	0.0014
All US dams (average power divided by collection area)	0.0049
Typical wind farm (with prevailing winds)	4
Typical wind farm (with winds from random directions)	1.2
Solar Two	3.2

Can the "solution" solve the problem? Table A11 shows the power consumption (all types of energy) in EJ/year and in TW_t from various sources. Roughly speaking, demand is increasing at 1 percent per year. Also note that a typical large power plant produces about 1,000,000,000 W_e.

Table A11: Average [thermal] power
consumption (1998) (by source)

	exajoules/year (EJ/year)	terawatts (TW$_t$)
United States	101	3.2
World	402	12.8
US petroleum	38.6	1.22
US coal	22.8	0.722
US natural gas	23	0.729
US nuclear	7.55	0.238
US hydro	3.79	0.120
US biomass (includes farm & timber waste)	3.22	0.102
US geothermal	0.332	0.0105
US solar	0.078	0.0025
US wind	0.038	0.0012

Electricity is clearly the most versatile form of energy, but there are no electricity wells. Table A12 shows the electricity picture in the US, both in billion kWh produced in 1998, and in around-the-clock average production in GW$_e$ (= million kW$_e$ = 10^9 W$_e$). Wind and solar combined, produced 0.12% of our electricity — one part in 826. They produced 0.6% as much electricity as our much maligned nuclear power plants — one part in 600.

Table A12: Average Electrical Production 1998 (US, by source)

	billion kWh$_e$	Average production (GW$_e$, gigawatts)
Total	3620	412.9
Utilities	3212	366.4
Non-utilities	407	46.4
Coal	1872.2	213.5
Nuclear	673.7	76.8
Natural gas	532.0	60.7
Other gas	12.8	1.46
Conventional Hydro	328.6	37.5
Geothermal	13.9	1.59
Wood	33.6	3.83
Waste	21.1	2.47
Wind	3.5	0.40 see footnote[73]
Solar (thermo-elec. & PV)	0.9	0.102

Wind is the fastest-growing energy source in the US, but only if you look at percentages instead of capacity. The growth from July 1998 to July 1999 (see page 34) was 565 MWe *capacity* (perhaps 200 MWe around-the-clock average).

Aside from hydropower and biomass, solar power, whether from heat, photovoltaics has a capacity factor of about 15 to 20 percent. That is, when advocates tell us the capacity, we should divide by five or six to get the expected annual average power. The capacity factor for wind can, in principle, be pretty much anything from 0% to 90%, depending upon the size of the generator fitted to the turbine. Typically, the capacity factor for wind generating systems (in 2001) is designed to be about 35%.

The capacity factors for much-maligned nuclear power are given in Figure 50. The lowest value there is for Brazil, just over 50%. Capacity

---

[73] In July 1999, the US wind capacity was 2,455 MW, amounting to about 500 MW (= 0.5 GW) average power.

factors in some countries (and even in some utilities of the heavily regulated US) exceed 90%. The worldwide average is 71.2%; that for industrialized countries is 80.4%.

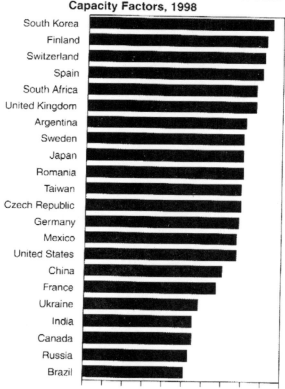

Figure 65. National Average Nuclear Power Plant Capacity Factors, 1998

Figure 50: Nuclear capacity factors throughout the world. The worldwide average capacity factor is 71.9%, with that of most countries exceeding 80%. Because of improved management, nuclear capacity factors in many US nuclear facilities are in excess of 90%.

Appendix B

Heat Engines

A heat engine is any device that can convert heat to mechanical energy. Examples are steam engines, gasoline engines, diesel engines, and natural-gas turbines. In all cases, an expanding gas does work by pushing something, either pistons or turbine blades. But that gas could not expand if everything in its environment were at the same temperature. That is, the gas must be hotter than its surroundings.

There is an upper limit to the efficiency of a heat engine that depends only upon temperatures. That is, if we could make an engine that had no friction whatsoever and which wasted no energy whatsoever, it would still be impossible for the efficiency to be 100%. The reason lies in our surroundings. The ambient temperature is not absolute zero.

Another limitation on the efficiency of heat engines comes from the materials with which we build them. Higher efficiency would be possible if we could allow the temperature to rise up to, say, the temperature of molten iron. But that would destroy our engine.

The limiting efficiency of heat engines thus depends upon the high temperature of the expanding gas T_{Hi} and the low temperature of the surroundings T_{Lo}. The equation for the maximum thermodynamic efficiency is

$$\eta_{max} = \frac{T_{Hi} - T_{Lo}}{T_{Hi}} = 1 - \frac{T_{Lo}}{T_{Hi}} \qquad \text{Eq. 1}$$

wherein all temperatures are measured on the absolute scale. The temperature of our surroundings is about 300 K (kelvins) and the temperature of the expanding gas in heat engines is typically about 500 K to 550 K. For $T_{Hi} = 500$ K, the maximum possible efficiency would be 1–300/500 = 0.4, or 40%.

Insulation and R-value

The rate of heat flow through a wall depends upon the thickness of the wall, the area of the wall, the temperature difference across the wall, and the properties of the material that makes up the wall. (See Fig. 51.) The following two paragraphs are details for readers who may wish to understand the *R*-value of insulation used by manufacturers of insulation.

In practice, wall insulation is made up to certain thicknesses, so that one can define a *thermal resistance R*, or *R*-value to the insulation. Letting *A* represent the area of the wall, we have a simple equation,

$$\text{Heat Flow Rate} = \frac{A(T_{Hi} - T_{Lo})}{R} \qquad \text{Eq. 2}$$

Unfortunately, the *R*-value for commercial insulation is usually given in the British system of units. To use the equation, multiply the wall area in square feet by the temperature difference in degrees Fahrenheit. Divide by the commercial *R*-value. The answer is the heat flow rate in BTU per hour. Multiply that answer by 0.293 to get the heat flow rate in watts.

For example, the heat flow rate across a 200-ft^2 wall with an R-value of 15 when the interior temperature is 70°F and the exterior temperature is 10° is (200)*(70 − 10)/15 = 800 BTU/hour, or 234 watts.

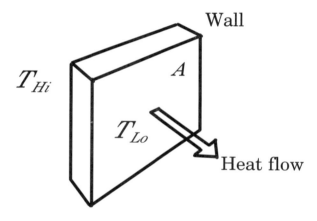

Figure 51: Heat flow through a wall. The larger the area *A* of the wall, the more readily the heat flows; ditto for the temperature difference $T_{Hi} - T_{Lo}$ across the wall. The greater the thickness of the wall, the less easily the heat flows. Finally, the material of the wall has an effect as well. Good insulators restrict heat flow.

The important things to remember about insulation are that the heat flow rate increases with increasing temperature difference across the wall and with increasing area of the wall. The heat flow rate decreases as the wall gets thicker. We have not discussed air flow through a wall, but obviously the less air flow, the less heat the air can carry.

Equation for Wind Power

For the benefit of interested readers, we derive Eq. 1 on page 114 for the power available from the wind.

Consider a column of air moving toward a wind turbine. It has cross-sectional area A and length ΔL (where the Greek letter Δ tells us to consider some small, arbitrary length), and therefore volume ΔV. If the density of the air is represented by ρ, then the mass in the column of air is $\rho \Delta V$. The air moves with velocity v, so its kinetic energy ΔE is $\Delta E = (1/2)\Delta M v^2 = (1/2)\rho \Delta V v^2 = (1/2)\rho A \Delta L v^2$. We will define ΔL to be the length of air column that reaches the wind turbine in a short time interval Δt because of its speed v. Therefore, we can write the energy of the column as $\Delta E = (1/2)\rho A v \Delta t \times v^2 = (1/2)\rho A v^3 \Delta t$. Now we have an equation for the kinetic energy of an arbitrary amount of air. Big deal. We really care instead about the rate of arrival of energy, $\Delta E / \Delta t$, which is the power P_w in the wind. That is, $P_w = (1/2)\rho A v^3$.

The density of the wind can most easily be found from the knowledge that one mole of gas occupies 22.4 liters at standard temperature and pressure. There are, of course, two main gases in air, nitrogen (80%) and oxygen (20%). The mass in the 22.4 liters is 0.8*28 grams + 0.2*32 grams = 28.8 grams. The density ρ is therefore 22.4 grams divided by 22.4×10^{-3} m³, which is 1.29 kg/m³.

We will assume that the fan of the wind turbine is a circle of radius R. The area is therefore $A = \pi R^2$

The power output P_{out} of the wind turbine is the wind power P_w multiplied by the efficiency η. Therefore, we have

$$P_{out} = (1/2)\eta(1.29)\pi R^2 v^3 = 2.02\eta R^2 v^3 \qquad \text{Eq. 3}$$

Any attempt at precision would be out of place here, as indeed it is when wind turbines are in operation, so we will round the 2.02 off to 2.

People who build wind turbine blades care how long they are, that is, the radius of the wind turbine; people who look at wind turbines are more likely to think of the diameter. We therefore give two equations for the wind turbine power, one using the radius, one using the diameter.

$$P_{out} = 2\eta R^2 v^3$$
$$= 0.5\eta D^2 v^3$$

The efficiency of an excellent wind turbine under ideal conditions will be about 50%, although most three-bladed wind turbines get up to only 45%. Moreover, in typical usage, the wind turbine achieves this power only at one wind speed. At higher speeds the efficiency is deliberately reduced by feathering the turbine blades. Using the 50% figure, we can arrive at a rule-of-thumb for the maximum power that an excellent wind turbine can achieve.

$$P_{out} = R^2 v^3$$
$$= 0.25 D^2 v^3$$

Wind-site Predictions

Our web site gets many questions from people who are interested in using wind turbines, but who have no idea of the many subtleties that are involved. The most important consideration is the distribution of wind speed at a given site. That information can be used to predict performance, as we show in this section.

The wind speed is never, say, 4.667221236825... meters per second. Therefore, the procedure is to round off wind speeds to the nearest 0.5 m/s. Figure 5 shows data taken from a wind farm in Kotzebue, located in a remote region of Alaska, just north the Arctic Circle. There were (for example) 490 hours during 1999-2000 when the wind speed was between 5.75 and 6.25 meters per second. Similarly, there were 585 hours when the wind speed was less than 0.25 m/s.

Usually, however, the data are more regular than the one shown in Fig. 52. The Rayleigh distribution usually matches the data better than it does in this unusual case; however, the bad fit at low windspeed in Fig. 52 isn't particularly important because at those speeds, it makes precious little difference whether the turbine turns or not.

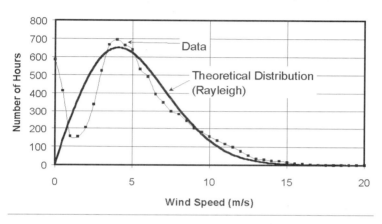

Figure 52: The wind speed at Kotzebue, Alaska (squares) and a simplified theoretical (Rayleigh) representation (heavy line).

The Rayleigh distribution[74] in Fig. 52 has a general shape, but the curve can be adjusted to lower or higher average wind speeds by adjusting one parameter V, as shown in Fig. 52. The value of V is $V = 1.13V_{avg} = \sqrt{2}V_{peak}$, where $V\text{-}_{peak}$ is the speed at which the curve reaches its maximum value.

---

[74] The quantity $(1/V^2)v\exp(v/V)^2 dv$, where V is a reference velocity, gives the probability that the velocity lies between v and $v + dv$.

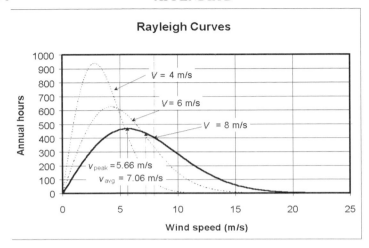

Figure 53: Rayleigh curves for various reference wind speeds V.

The next thing to do is to predict the annual amount of energy that can be obtained from the wind. The answer, of course, depends upon the size and quality of the wind turbine. Let us assume that the Rayleigh curve that fits our wind speed has a value of $V = 8$ m/s. We have a wind turbine that is capable of producing 100 kW$_e$ when the wind speed is 11.5 m/s, and that (by design) the wind turbine will produce that power for all wind speeds up to 25 m/s, after which it will be turned off.

There are 199 hours during the year when the wind speed is 11.5 m/s (give or take 0.25 m/s), so that will contribute 19,900 kWh of energy to the annual total. Similarly, there are 287 hours a year when the wind speed is 10 m/s ($\pm$ 0.25 m/s); however, the output power will be less than 100 kW$_e$. In fact, it will be 100 kW$_e$ times the cube of (10/11.5), or 66 kW$_e$. Those 287 hours will therefore contribute 18942 kWh to the annual total. To find the total annual energy output, one repeats the procedure for every wind speed.

Given the same wind turbine, we could equip it with a 200-kW$_e$ generator (which produces full power at 14.5 m/s) or a huge generator capable of unlimited power. See Fig. 4 to see the power output from each of the generators. We would then proceed to do the same calculations to find the annual energy output.

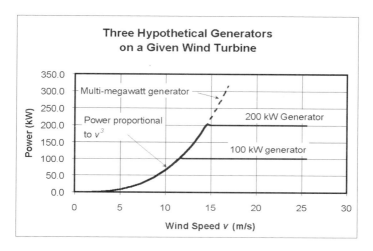

Figure 54: Power from one hypothetical wind turbine, equipped with either a 100-kW$_e$ generator, a 200-kW$_e$ generator, or a generator of unlimited power output.

The outputs of the generators are shown in Fig. 55 for various wind speeds. The shaded area under the curve represents the total energy output for a year for the 100-kW$_e$ generator. The unshaded area that lies above the 100-kW$_e$ curve and below the 200-kW$_e$ curve represents the annual difference between the energy production of the 200-kW$_e$ generator and the 100-kW unit, given the same wind turbine.

The annual output of the 100-kW$_e$ generator is 296,000 kWh, and that of the 200-kW$_e$ generator is 359,000 kWh. That is, the 200-kW$_e$ generator produces only 21% more energy in a year than the 100-kW$_e$ generator. The capacity factor — the ratio of average power to nameplate power — of the larger generator is 20%, but the capacity factor of the smaller one is 34%. If there were a 10-MWe generator attached, it would produce 392,000 kWh in a year, 32% more than with the 100-kW$_e$ generator. The capacity factor would be 0.0004%.

It is extremely important to understand the wind speed at a given site. In our model above, if the reference wind speed were 7.5 m/s instead of 8 m/s, the annual energy for the 100-kW$_e$ generator would be 12% lower (261,000 kWh), and the capacity factor would be 30% instead of 34%.

The story doesn't stop there. From Fig. 53, it is apparent that even for the 100-kW$_e$ generator, *most* of the time — 86% for this example — the wind speed is too low (*i.e.*, below 11.5 m/s) to produce full power. The 34% power factor says much the same: most of the time, the power production is well below the nameplate power rating of the generator, even though there is a wide range of windspeed — 11.5 m/s to 25 m/s — where full power is produced

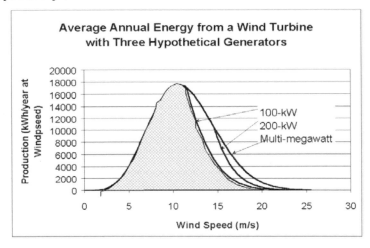

Figure 55: The electrical output in kWh for a wind turbine equipped with either a 100-kW$_e$, a 200-kW$_e$, or a multi-megawatt generator, for various wind speeds. Area represents annual energy. (The bumps and wiggles are artifacts of the drawing program.)

Whenever the windspeed is below 11.5 m/s — that is, *most* of the time — the power output is not constant, but is proportional to the cube of the windspeed. This regime is the curved portion of the power curve in Fig. 54. If the wind speed fluctuates, then the output power fluctuates. When the wind speed is low, the power fluctuations have little effect on the power line. however, when the wind speed is higher, so that the output power is between 50 kW$_e$ and 100 kW$_e$, then the fluctuations require the output of conventional power stations to rise or fall in response. It is of interest that the output power is between half-power and full power about 15% of the time.

Equation for Water Power

Water has potential energy by virtue of being above some reference point. For a given incremental mass Δm of water, its potential energy ΔPE is $\Delta PE = \Delta mgh$, where g is the acceleration due to gravity, 9.8 meters per second per second (9.8 m/s^s), and h is the elevation, called the *hydraulic head* in Fig. 24 and subsequent text. The power (of the water) is the potential energy loss ΔPE divided by the time interval Δt during which the water moves down in elevation by h. That is, $P_{water} = \Delta PE/\Delta t$. Usually, however, one expresses the quantity of water in volume units, not mass units. The density ρ of water is 1000 kg/m^3. We have

$$P_{water} = \frac{\Delta PE}{\Delta t} = \frac{(\Delta m)gh}{\Delta t} = \frac{\rho gh \Delta V}{\Delta t} = \rho gh \frac{\Delta V}{\Delta t} \qquad \text{Eq. 4}$$

where $\Delta V/\Delta t$ = the rate of flow of water in cubic meters per second. The power output of the plant is the power of the water multiplied by the efficiency, η, which we take to be 85%. Inserting the numbers, we have

$$P_{out}(\text{watts}) = 8330 \times h(\text{meters}) \times \frac{\Delta V}{\Delta t}(\text{cubic meters per second})$$

In "Questions about Hydropower" page 63, we asked why water falling over a dam doesn't boil. If one kilogram of water descends 100 meters (about 300 feet) through a hydropower station, it generates enough energy to raise the temperature of one kilogram of water only about 0.23 °C. To raise one kilogram of water from room temperature (20 °C) to the boiling point (100 °C) requires about 343 kg of water to descend through that 100-m hydropower station.

PV cells as diodes

Photovoltaic cells are, in fact, rather amazing. But there's no magic involved. The devices have properties that can be measured by anybody who takes the time.

PV cells are made of dissimilar semiconductors placed in contact. So are solid-state diodes, devices that conduct current one direction but not the other. In fact, PV cells are diodes and that statement has implications.

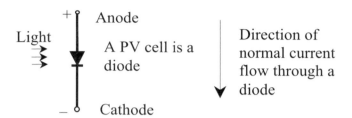

Figure 56: A PV cell is a diode. When light shines on the cell, the anode becomes positive, *against* the normal direction.

Diodes are one-way devices for current, which flows in the direction of the arrow (see Fig. 56). That is, the current through the diode flows easily from anode — a p-type semiconductor — to cathode — an n-type conductor; current from cathode to anode is blocked.

When light shines on a PV cell, a voltage is generated such that the anode is positive. That is, current goes "backwards" through the diode, the direction it would resist conducting if a battery were attached to make it do so. The tendency for the current to leak back through the diode in its normal direction of flow — from the positive anode to the negative cathode — is real. In fact, the term *recombination current* is used to describe that very phenomenon.

It is important here to recognize that a PV cell actually generates a *current*, not a voltage. Although there is a voltage present, it is determined by something other than sunlight shining on the PV cell.

Here is what happens: sunlight dislodges electrons from their atoms, leaving positive ions behind. The positive spaces are called *holes*. As electrons hop from a neutral atom to a hole, leaving a hole behind, it appears as if the holes themselves are moving. This is the nature of the current in a p-type semiconductor. Sunlight produces *electron-hole* pairs; the electron goes one way, the hole goes the opposite, resulting in the p-type semiconductor becoming positive and the n-type becoming negative.

Diode Behavior of PV Cell

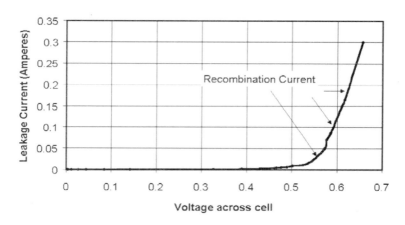

Figure 57: The current-voltage characteristics of a silicon PV cell.

Fortunately, the PV cell is a very non-linear device, as shown in Fig. 57. It conducts very little recombination current when the voltage is low, less than about 0.5 volts for the example shown in Fig. 57.

Suppose that the PV cell of Fig. 57 can generate 100 milliamperes when exposed to full sunlight. If the cell is not connected to a load, the generated current would exactly equal the recombination current; by Fig. 57, the voltage would be just under 0.6 volts. The efficiency would be zero, owing to the fact that (by design) no power is being delivered to do a load.

On the other hand, if we connected the PV cell to a load, some of the generated current would go through the load, and much less through the PV cell itself. For example, the voltage might drop to 0.5 volts. The recombination current (Fig. 57) would be 10 milliamperes, and the lion's share of the current — the other 90 milliamperes — would go through the load. The trick to using PV cells efficiently is to keep the voltage low enough that the recombination current is negligible.

In other words, the efficiency of a PV device depends upon the load to which it is attached. The actual behavior of a PV array will be discussed in figures 59 and 60.

Birdie, Birdie in the Sky

The diode characteristics have other implications for PV cells. Because they inherently generate low voltage, PV cells are normally placed in series to add the voltages, as shown in Fig. 58. It could conceivably happen, possibly through the courtesy of a bird, that one or more PV cells could be shaded. The light-generated current is always going "backwards" through the diodes, but when a PV cell is shaded, it generates no current and merely blocks that passage of current generated by the other cells. Fortunately, the PV cell is not a good diode; the voltage need not be very high before it will permit current to pass through it the wrong way.

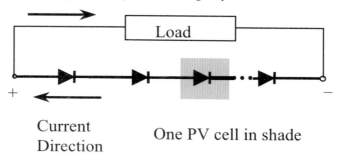

Current
Direction

One PV cell in shade

Figure 58: A PV array with one PV cell in shade.

I borrowed an 18-cell PV array from the student labs at the University of Southern Colorado. (The cells are amorphous silicon, cheaper but not as efficient as crystalline silicon wafers.) The characteristics of the device when connected to an adjustable load are shown in the upper curve in Fig. 59. When the load resistance is high, the current is very low and the voltage is high, as shown at the lower right of Fig. 59. The power — the product of high voltage (22V) and zero current (0) — is zero. That is, the efficiency is zero (or near zero) when either the PV cell is connected to a load resistance that is either too high or too low. The maximum efficiency occurs when the load resistance is such that the "operating point" is near the knee of the curve of Fig. 59, as shown by the tip of the arrow.

PV output

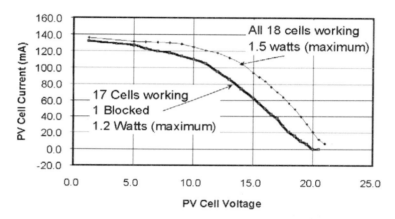

Fig. 59: The output from a PV array with all 18 cells exposed to full sunlight (upper curve) and with one cell shaded.

When the load resistance is very low, the effect of sunlight is merely to drive electrons around a loop that does not inhibit the flow. The current is high (136 mA), but the voltage is zero, so, again, the power is zero. When everything is adjusted just right, the power is maximized at about 13.6 volts, 110 mA, (tip of arrow) where the output power is 1.5 watts. The efficiency of the unit is about 5%.

The lower curve in Fig. 59 shows what happens when one of the 18 cells is shaded by a piece of wood placed over the cell. In this case, the maximum power output is reduced from 1.5 watts to 1.2 watts, a 20% decrease caused by a 5.6% reduction in the number of cells exposed to sunlight. The efficiency of the 18-cell array, that is, has dropped from 5% to 4% because of that one blocked cell.

Power to the People!

It is normal for the delivered power to depend upon the voltage source, whether the source be hydropower, coal, or other fuel. If a customer demands more power, the demand is immediately felt at the generator, where the rotation rate diminishes. Immediately, the generator is brought back up to speed by the application of more steam, more natural gas (turbines), or more water (for hydro).

When the demand for PV power increases, however, the PV cell can do nothing to help, short of begging the sun for more light. Of course, the power output will *decrease*, because the PV array will already be operating at maximum power, solar power being so feeble to begin with.

PV (Amorphous Silicon) Output
(18 Cells in Series)

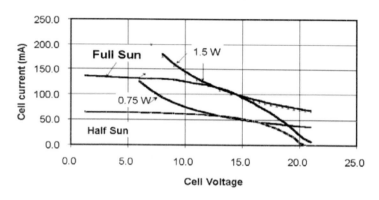

Figure 60: The output from an 18-cell PV array in full sunlight (upper) and half sunlight (lower). The curves labeled 1.5 W and 0.75 W represent constant power: $P = VI = 1.5$ watts and $P = IV = 0.75$ watts.

"Rooftop systems that can meet half a home's electricity needs now cost as little as $10,000 with rebates and tax credits from the federal and state governments."

The Associated Press 8/07/2001.

Demand Instability

Suppose we consider theoretically the equation $PV = 1.5$ watts. That is, we ask what the voltage would be for each current we might consider, such that the power delivered to the load would be 1.5 watts. Algebraically, the curve is just a hyperbola when I is plotted against V, as it is for any given power. We can superimpose the hyperbolic curve for 1.5 W on a graph showing real data from our 18-cell array, as shown in Fig. 60. Just below the 1.5-watt curve are some tiny crosses representing 1.4 watts.

The region where the 1.5-W curve touches the full-sun curve shows a small region where the power output is nearly constant. (Actually, there is only one point of contact.)

The curve representing 1.4 watts is shown as the little crosses in Fig. 60. You can see that if the demand for power increases — which always shows up as a demand for more current, the output of the PV cell would move upward to the left, soon intersecting the 1.4-W curve. To repeat the point, increased demand for power results in *decreased power output from a PV array*. The effect is to decrease the efficiency at the same time.

For the 18-cell array producing its maximum power, a 10% increase in current results in a 20% decrease in voltage, therefore a 10% decrease in power. A demand for increased power results in a decreased supply of power, resulting in a stronger demand for power, resulting in even less power. Overall, then, there is *demand instability*. When more power is demanded, the PV cell can't supply it; worse yet, the PV power *decreases*.

This is, in a way, what happened with the Great Northeast Blackout of 1965. With no ability to compensate for increased demand (due to the malfunction of one relay), the system responded by reducing power (circuit breakers removing power plants from the grid).

Less Light, Less Power

Figure 60 shows the output of the 18-cell array in full Pueblo, Colorado, sunlight and in half sunlight, as one might have on a hazy day (one where the sun appears to be somewhat lost in haze, but there are still distinct shadows). Half sunlight can also occur if the array is 60° from facing the sun directly (morning conditions). The largest effect of half-sun conditions is a reduction of the current at which maximum power occurs; the optimum voltage is basically unaffected.

The terms *solar cell* and *solar battery* give the wrong impression. Unlike batteries, PV collectors do not store energy. The instant the PV collector is shaded, it stops producing electricity.[75] It is important to recognize that if PV cells are to provide electricity when the sun is not shining, the energy must be stored, somewhere, somehow.

---

[75] For that matter, it is useful to remember that the electrical grid is not a storage device. At all times, the electricity being generated is being used somewhere.

References

Annual Energy Outlook 1998 (Energy Information Agency of the Department of Energy, available from US Government Printing Office.)

APS report: *Solar Photovoltaic Energy Conversion: Principal Conclusions of the American Physical Society Study Group on,* (American Physical Society, January, 1979), Ehrenreich, H., Chairman

Beckmann, Petr, *Access to Energy,* (Oct. 1978,

Berman, Daniel M. and John T. O'Connor, WHO OWNS THE SUN?: *People, Politics, and the Struggle for a Solar Economy* (Chelsea Green Publishing Company, White River, VT 1996).

Blackburn, John O., THE RENEWABLE ENERGY ALTERNATIVE: *How the United States and the World Can Prosper Without Nuclear Energy or Coal* (Duke University Press, 1987).

Bretschneider, C. L., "Sea Motion," in *Handbook of Ocean and Underwater Engineering,* Myers, John J., Carl H. Holm, editors, (McGraw Hill, New York, 1969).

Bryan, Ford R. *Henry's Attic,* (Ford Books, Dearborn, MI, 1995).

Brower, Michael, COOL ENERGY: *Renewable Solutions to Environmental Problems,* (The MIT Press, Cambridge, MA, 1992).

Brown, Lester R., Christopher Flavin, and Sandra Postel, SAVING THE PLANET: *How to Shape an Environmentally Sustainable Global Economy* (W.W. Norton, New York, 1991).

Bryson, John E. (President, Calif. PUC), FINANCING THE SOLAR TRANSITION: *A Report To The California State Legislature*, (Calif. PUC Jan 2, 1980).

Commoner, Barry, Howard Bokensenbaum, and Michael Corr, eds. ENERGY AND HUMAN WELFARE—A CRITICAL ANALYSIS, *Volume II: Alternative Technologies for Power Production,* (Macmillan Information, New York, 1975).

Commoner, Barry, "Prepared Statement of Barry Commoner" for Birch Bayh's hearings on biofuels, *Alcohol Fuels Hearings* (1978).

Considine, Douglas M., P.E., Editor, *Energy Technology Handbook,* (McGraw-Hill, 1977).

Ehrenreich, H., Chairman, *Solar Photovoltaic Energy Conversion: Principal Conclusions of the American Physical Society Study Group on,* (American Physical Society, January, 1979).

Ehrlich, Paul R. and Anne H. Ehrlich, THE END OF AFFLUENCE, *A Blueprint for Your Future*, (Rivercity Press,1974).

Ehrlich, Paul R. & Anne H. Erhlich, HEALING THE PLANET: *Strategies for Resolving the Environmental Crisis* (Center for Conservation Biology, Stanford University, 1991).

Energy and Power, Scientific American offprint book, W.H. Freeman (1971).

Hayes, Denis RAYS OF HOPE: *The Transition To A Post Petroleum World* (W. W. Norton & Company, 1977).

Hubbard, H. M., "Photovoltaics Today and Tomorrow," *Science* **244**, pp.297–304 (21 April 1989).

Huff, Darrell, *How to Lie with Statistics*, (W.W. Norton, Inc., New York, 1954).

International Energy Outlook 2001 (Energy Information Agency of the Department of Energy, available from US Government Printing Office.)

Jackson, Barbara Ward and René Dubos ONLY ONE EARTH: *The Care and Maintenance of a Small Planet*, (W.W. Norton & Company, New York, 1972).

Keyes, John, THE SOLAR CONSPIRACY: *The $3,000,000,000,000 game plan of the energy barons' shadow government*, (Morgan and Morgan Publishers, Dobbs Ferry, NY, 1975).

Lovins, Amory, SOFT ENERGY PATHS: *Toward a More Durable Peace* (Balinger Publishing Co., Cambridge, MA, 1977).

Lovins, Amory, "Tough Lovins," *The Weekly Standard,* p. 6, (June 4, 2001).

Meinel, Aden E. and Marjorie P. Meinel, *Applied Solar Energy,* (Addison-Wesley Publishing Co., Reading, MA, 1976)

Naar, Jon, DESIGN FOR A LIVABLE PLANET, (Harper and Row, 1990).

Nader, Ralph, 1997: *Frontline* (PBS Broadcasting, 4/22/1997).

Oppenheimer, Michael, and Robert Boyle, DEAD HEAT: *The Race Against the Greenhouse Effect* (Basic Books, New York, 1990).

Reese, Ray, THE SUN BETRAYED: *A Report on the Corporate Seizure of U.S. Solar Energy Development,* (South End Press, P.O. Box 68, Astor Station, Boston, 1979).

San Pietro, Greece, and Army, eds., *Harvesting the Sun: Photosynthesis in Plant Life*, (Academic Press, New York, 1967).

Robert F. Service, "A Record in Converting Photons to Fuel," *Science* **280**, p. 382 (17 April, 1998).

Statistical Abstract of the United States (available from US Government Printing Office.)

Train, Russell, Chairman, CHOOSING A SUSTAINABLE FUTURE: *The Report Of The National Commission On The Environment,* (Island Press, Washington, D.C. 1993).

C. Wu, "Power Plants: Algae churn out hydrogen," *Science News*, p. 134 (Feb. 26, 2000).

Web Sites

http://www.cato.org/pubs/pas/pa-241.html

http://www.commondreams.org/views01/0708-05.htm, Amory B & Hunter L. Lovins, "Too Expensive and Unacceptably Risky, Nuclear Power was Declared Dead Long Ago. So Why Would We Resurrect It?" (Dec. 10, 2001).

http://www.eia.doe.gov

http://www.energyadvocate.com

http://www.energy.ca.gov/wind/windfacts.html

http://www.energy.ca.gov/wind/wind-html/95_wind_report.html

http://www.epa.gov/globalwarming/publications/actions/state/wa/mitigatef.html

Browner 1998: http://www.epa.gov/oppeooe1/globalwarming/actions/clean-energy/sol/browner_498.html

http://www.es.wapa.gov/pubs/esb/97Oct/at_roof.htm

http://www.greenpeaceusa.org/media/factsheets/windtext.htm

http://www.jxj.com/magsandj/rew/1999_04/comingofage.html (Christopher Flavin & Seth Dunn, 1999)

http://teamhouse.tni.net/janebio.htm (Jane Fonda, 2000)

http://www.nrel.gov

http://www.nrel.gov/data/pix/searchpix.html

http://www.nrel.gov/documents/profiles.html

http://rredc.nrel.gov/solar/#archived.

http://people.cornell.edu/pages/tg21/usgs.html

http://www.rmi.org/sitepages/pid189.php (Amory Lovins, 2001)

http://www.tidalelectric.com/events/archives/archive0298.html (Gore $48 billion)

Worldwatch (1996): http://www.worldwatch.org/alerts/pr960814.html

http://www.worldwatch.org/alerts/010517.html (Hydrogen as *source* of energy)

Glossary

Many terms are defined throughout the book. It may be useful to consult the index or the tables in Appendix A for terms not listed here.

/	Divided by, or per. For example J/s means joules divided by seconds or joules per second.
AC	Alternating current.
ampere	The unit used for the rate of flow of electric charge, equal to one coulomb per second, or about 6.28×10^{18} electrons per second.
availability factor	A factor telling what percentage of the time a system is ready to be used if it is needed.
barrel	42 gallons (US) = 0.1590 m³ Quantity often used to describe quantity of oil; not to be confused with *drum* (55 gallons)
BTU	British Thermal Unit, the amount of energy required to raise the temperature of one pound of water by one degree Fahrenheit.
calorie	The amount of energy required to raise the temperature of one gram of water by one degree Celsius. Notice the lower-case c. (This is not a food Calorie.)
Calorie	The amount of energy required to raise the temperature of one kilogram of water by one degree Celsius. A food Calorie — often written incorrectly with a small c — is a Calorie.
capacity	The maximum power output of a generator of generating station.
capacity factor	The ratio of the energy produced by a power plant in a year to the energy it could have produced if it ran at full power for the year, often expressed as a percentage.
charge	Excess or deficiency of electrons, usually measured in amperes × seconds, *a.k.a.* Coulombs
Coulomb	1 ampere-second
current	The rate of flow of electric charge. Unit: ampere
DC	Direct current. (Means unidirectional current, but not necessarily steady current.)

DOE	Department of Energy
efficiency	The ratio of the energy delivered to the energy input, often expressed as a percentage.
EIA	Energy Information Agency (of the DOE)
energy intensity	Energy per unit time per unit area, the same units as those of solar intensity. Unit: W/m^2
environmentalist	(as used in this book): a person who rises to object to other people's activities on the grounds of defending the environment
ethanol	Ethyl alcohol, CH_3CH_2OH
EtOH	Ethanol.
flux, solar	Solar intensity. Unit: $J/s/m^2 = W/m^2$
generator	A rotating machine that converts mechanical energy into electrical energy.
GW	Gigawatts ($= 10^9$ watts)
GW_e	Gigawatts, electric. Distinguished from GW_t, gigawatts, thermal
head	See hydraulic head.
heat content	The amount of energy per unit of mass that can be obtained from burning a fuel. Unit: J/kg, joules/kilogram
heat engine	A mechanical device for converting heat to work. Examples are steam engines, gasoline engines, and gas turbines.
hectare	10,000 m^2. An 100-m^2 plot of land was considered suitable for a garden, and was called an *are*. One hundred of them comprise a hectare.
horsepower	A common non-SI unit of power, Abbrev: HP, equal to about 746 watts.
HV	High voltage.
hydraulic head	The elevation of the water in a reservoir above the lower river to which it is discharged. Available hydropower is greater if the hydraulic head is greater.
hydropower	Power obtained from water as it descends to a lower elevation.
insolation	solar intensity

intensity, solar	Energy per unit area per unit time. Unit: $J/s/m^2 = W/m^2$
J	Abbreviation for joule
joule	SI unit of energy. Abbrev: J. 1 J = 1 W•s (watt-second)
kW_e	kilowatts, electric
kWh	kilowatt-hour. (1 thousand watt-hours = 3.6×10^6 J)
langley	One calorie/cm^2 (1 langley/day = 0.484 W/m^2)
MW_e	megawatts, electric
MWh	megawatt-hour (1 million watt-hours = 3.6×10^9 J)
MW_e	Megawatts, electric. Distinguished from MWt, megawatts, thermal
nameplate	also nameplate power rating. The capacity (maximum power rating) of a generator or power station.
NREL	National Renewable Energy Laboratory (*nee* SERI)
ohm	Unit for electrical resistance. 1 ohm = 1 volt/ampere.
Ohm's Law	There is a class of devices for which the current I through them is directly proportional to the voltage V across them. That is, $I = \dfrac{V}{R}$, where R is the resistance of the device.
optical efficiency	The ratio of the sunlight delivered to a target by a mirror or lens to the incident sunlight on that mirror or lens.
per	Divided by; apiece
photovoltaics	The conversion of sunlight directly into electricity through the use of solar cells.
power	Energy per unit time.
PV	See photovoltaics
quad	One quadrillion (10^{15}) BTU.
R-value	Thermal resistance, usually expressed in non-SI units. See Appendix B.
RPM	Revolutions per minute. (The SI unit is radians per second.)
radian	An angle whose arc length equals one radius. There are 2π radians in one full revolution.

renewable	capable of being renewed by natural ecological cycles sufficiently abundant that the supply can be considered infinite (such as geothermal)
SERI	Solar Energy Research Institute (now NREL)
SI	*Systéme Internationale* (International system of units)
solar cells	Semiconductor devices that absorb sunlight and produce electricity.
technosolar	Any solar-energy collection scheme that requires direct interception of sunlight by manufactured equipment.
thermal resistance	Resistance to the flow of heat. See Insulation and R-value, p. 187
ton	Two thousand pounds.
tonne	One thousand kilograms. (=2205 pounds, about 10% greater than a ton.)
transformer	A device, made with coils of wire, for converting low AC voltage to high AC voltage, or conversely. It does not create energy. The input power $V_{in}I_{in}$ equals the output power $V_{out}I_{out}$ if the efficiency is 100%. (Most transformers are over 98% efficient.)
turbine	A rotary engine with blades that are caused to rotate by (usually) wind, water, steam, or burning natural gas.
voltage	The energy per unit charge. 1 volt = 1 joule/(ampere-second)
watt	SI unit of power. Abbrev: W (1 W = 1 J/s, joule/second). For electrical power, 1 watt = 1 volt × 1 ampere (if the current is in phase with the voltage)
work	A measure of effort expended, calculated from the applied force multiplied by the distance moved.

Index

About the Author

Howard Hayden is a Professor Emeritus of Physics from the University of Connecticut and Adjunct Professor at the University of Southern Colorado.

A Colorado native, he entered the University of Denver as an engineering major, but soon discovered that he wasn't temperamentally suited to all that reality. He switched to physics and went on to earn his B.S., M.S., and Ph.D.

On receiving his Ph.D., he went to the University of Connecticut where he spent 32 years clasting icons and corrupting young minds. He did accelerator-based atomic physics, including measurements of cross-sections for various processes, measurements of energy loss in atomic collisions and of lifetimes of excited states, beam-foil spectroscopy, and ion implantation.

He also repeated the famous Trouton-Noble experiment, but with 100,000-times greater sensitivity. For nearly a century, it had been believed that the Trouton-Noble experiment contradicted classical physics and lent support to Einstein's special relativity theory, but Hayden proved that the experiment did not contradict either classical theory or Einstein theory. Therefore, the original experiment was essentially meaningless, and so was Hayden's. But nobody knew that until the experiment was reconsidered for the first time in a century.

Hayden has a long-standing interest in energy, stemming from before the OPEC oil embargoes of the 70s. Presently, he publishes a monthly newsletter, *The Energy Advocate*, now (2001) in its 6th year of publication.

THE ENERGY ADVOCATE

If you have found *The Solar Fraud* interesting and informative, you may be interested in THE ENERGY ADVOCATE, a monthly newsletter that promotes energy and technology.

Our monthly newsletter contains technical articles written for intelligent non-experts. Our subscribers are found in about a dozen foreign countries as well as throughout the US. Among our subscribers are engineers, physicists, chemists, economists, lawyers, geologists, corporate CEOs, high-school teachers, college and university professors, doctors, dentists, environmental scientists, military personnel, computer programmers, and accountants, to name a few.

We also have lively commentary about the political forces that oppose technology and that seek to stifle energy production. From the hooligans who blockade the entrances to power plants under construction to the privileged and pampered political leaders who try to destroy nuclear power plants, the coal industry, the petroleum industry, and the hydropower plants, the anti-technological forces are a pervasive component of modern society. The national news media, including (but not limited to) the *New York Times* and *CBS News* glorifies the obstructionists; we denounce them without remorse.

If THE ENERGY ADVOCATE carried advertising, we might have to compromise, to pull punches, to avoid offending our advertisers. But we maintain our independence by relying entirely on subscriptions. We are beholden to nobody but our readers.

The Taliban era is over, but medieval sheikdoms continue to rule the land that has the world's greatest supplies of petroleum. Keep up to date on energy topics. Send $35.00 to THE ENERGY ADVOCATE, P.O. Box 7595, Pueblo West, CO, 81007 for a one-year subscription.

DATE DUE

JUN 3 0 2003		
MAY 0 5 2004 DEC 1 7 2004		
NOV 2 6 2006		
2 2 4		